New Trends on Nonlinear Optics in Nanostructures and Plasmonics

New Trends on Nonlinear Optics in Nanostructures and Plasmonics

Editor

Gennady M. Mikheev

MDPI

Basel • Beijing • Wuhan • Barcelona • Belgrade • Novi Sad • Cluj • Manchester

Editor
Gennady M. Mikheev
Udmurt Federal Research Center
of the Ural Branch of the Russian
Academy of Sciences
Izhevsk
Russia

Editorial Office
MDPI
St. Alban-Anlage 66
4052 Basel, Switzerland

This is a reprint of articles from the Special Issue published online in the open access journal *Applied Sciences* (ISSN 2076-3417) (available at: https://www.mdpi.com/journal/applsci/special_issues/Nonlinear_Optics_nanostructures).

For citation purposes, cite each article independently as indicated on the article page online and as indicated below:

Lastname, A.A.; Lastname, B.B. Article Title. *Journal Name* **Year**, *Volume Number*, Page Range.

ISBN 978-3-0365-9370-8 (Hbk)
ISBN 978-3-0365-9371-5 (PDF)
doi.org/10.3390/books978-3-0365-9371-5

Cover image courtesy of Gennady M. Mikheev

Contents

About the Editor

Gennady M. Mikheev

Prof. Gennady M. Mikheev received the M.S. and Ph.D. from the Physic Department of Lomonosov Moscow State University in 1981 and 1985, respectively. He received the titles of Dr.Sci. and Prof. from the Russian National Examination Committee in 2000 and 2008, respectively. He is currently with the Institute of Mechanics, Udmurt Federal Research Center of the Ural Branch of the Russian Academy of Sciences, Izhevsk, Russia. His research interests include the interaction between laser radiation and matter, photonics, optoelectronics, and nonlinear optics.

Preface

Nonlinear optics is one of the most interesting and exciting areas of modern physics. It was established in 1961 with the discovery of ruby laser second-harmonic generation in crystalline quartz. In general, nonlinear optics involves studying the processes of interaction between electromagnetic radiation and matter, the behavior and features of which depend on the intensity of the incident radiation. The development of nonlinear optics is inseparably connected with the progress in the field of laser physics. Due to the availability of powerful lasers, it has become possible to produce an d observe very beautiful, and sometimes quite exotic, nonlinear optical effects. On the other hand, discoveries in the field of nonlinear optics made it possible to invent and build different lasers, operating in continuous, pulsed and pulsed-periodic modes in a wide spectral range of wavelengths, including the terahertz region, which has greatly enriched the scientific instruments of nonlinear optics itself. While the first works on nonlinear optics were carried out in bulk media using nanosecond lasers, the study of nonlinear optical properties of nanostructures and nanomaterials is currently popular to conduct using femtosecond lasers. Today, nonlinear optics continues to be a dynamically growing field of science due to the emergence of light sources with previously unattainable incredible characteristics and due to the impressive advances in nanotechnology that have provided researchers with nanomaterials that have fascinating properties. The study of the nonlinear optical properties of such materials provides new insights for the further development of photonics and optoelectronics.

The purpose of this Reprint is to inform readers about the findings of some of the experimental and theoretical works in the field of nonlinear optics carried out in the last three years in various laboratories around the world by research groups with high levels of expertise. The Reprint involves an Editorial and presents ten original studies from different areas of nonlinear optics. The object of most of these works is nonlinear optical phenomena in nanostructures and nanomaterials.

The Reprint may be useful for a wide range of specialists and researchers working in the field of laser–matter interaction, and may also be of interest to undergraduate and graduate students specializing in photonics.

Gennady M. Mikheev
Editor

applied sciences

MDPI

Editorial

Special Issue on New Trends in Nonlinear Optics in Nanostructures and Plasmonics

Gennady M. Mikheev

Institute of Mechanics, Udmurt Federal Research Center of the Ural Branch of the Russian Academy of Sciences, T. Baramzinoy Str., 34, 426067 Izhevsk, Russia; mikheev@udman.ru; Tel.: +7-3412-21-66-11

1. Introduction

Nonlinear optics, which emerged in the early 60s of the 20th century, immediately after the invention of powerful lasers, had a significant influence on the formation of modern photonics. Studies in the field of nonlinear optics led to such discoveries as the generation of second and higher harmonics of laser radiation, parametric scattering, stimulated Raman scattering, wavefront reversal, self-focusing and self-defocusing of light beams, and multiphoton absorption. The development of nonlinear optics, in turn, had a significant impact on the development of lasers and laser technologies. This resulted in the emergence of lasers based on stimulated light scattering, tunable lasers with a wide range of wavelengths, pico- and femtosecond lasers, and terahertz radiation generators, which significantly expanded the instrumental capabilities for research in the field of nonlinear optics. Fantastic results obtained in the field of nanotechnology over the last two decades have provided scientists with new, unique materials, the study of nonlinear optical properties of which was crowned with the discovery of new natural phenomena. Thus, at present, nonlinear optics is still a burgeoning branch of physics.

2. Summary of the Articles of the Issue

This Special Issue presents the results of studies conducted in various fields of nonlinear optics. These studies range from experimentally investigating the generation of higher harmonics of laser radiation in laser-induced plasma to theoretically examining the generation of edge currents in graphene caused by second-order nonlinearity. In [1], the peculiarities of the generation of the 17th and 18th harmonics of the femtosecond laser radiation (at a wavelength of 806 nm) in the laser plasma were studied. The plasma was excited in bulk Sn and Sn nanoparticles glued onto a glass substrate by powerful, focused radiation of picosecond and nanosecond lasers. Additionally, plasma probing was performed using bi-harmonic radiation at wavelengths of 806 and 403 nm (the second harmonic of a femtosecond laser) as well as the radiation of a femtosecond parametric light generator (1280–1440 nm) and its second harmonic in the single-frequency and double-frequency modes. It is shown that the use of a two-frequency mode allows achieving a 12-fold increase in the intensity of even harmonics. It was found that for the effective generation of higher harmonics pulses, it is necessary to provide an optimal delay (130–180 ns) between the plasma excitation laser pulses and pulses of probing radiation. The results of a study on the generation of higher harmonics in the wavelength range of 30–100 nm by a femtosecond laser at 806 nm in a laser-induced plasma from six carbon-containing materials (graphite, fullerenes, carbon nanotubes, carbon nanofibers, diamond nanoparticles, and graphene) are presented in [2]. It was demonstrated that the presence of nanoparticles (C_{60}, carbon nanofibers, and multi-walled carbon nanotubes) and their fragments in the laser plasma leads to an increase in the efficiency of 9–17th harmonic generation at wavelengths of 50–95 nm. The generation of laser radiation harmonics can be utilized to develop unique tools for studying the surfaces of materials. In [3], magnetization-induced second and third harmonic generation of a femtosecond laser radiation of an optical parametric oscillator at

Citation: Mikheev, G.M. Special Issue on New Trends in Nonlinear Optics in Nanostructures and Plasmonics. *Appl. Sci.* **2023**, *13*, 11601. https://doi.org/10.3390/app132011601

Received: 18 October 2023
Accepted: 21 October 2023
Published: 23 October 2023

740–1800 nm has been applied for the visualization of the interface magnetic domains of 10 μm thick $(LuBi)_3Fe_5O_{12}$ garnet film. It has been shown that the technique of nonlinear optical microscopy based on third harmonic generation provides better spatial resolution compared to microscopy based on second harmonic generation.

In the paper [4], comparative studies of the nonlinear optical properties of aqueous suspensions of high pressure-high temperature (HP-HT) nanodiamonds and detonation nanodiamonds under femtosecond laser excitation are performed. It was shown that at the same parameters of femtosecond laser pumping and experimental conditions, saturable absorption occurred in an aqueous suspension of detonation nanodiamonds accompanied by short-term nonlinear bleaching, while in an aqueous suspension of HP-HT nanodiamonds, two-photon absorption appeared, leading to optical limiting. In study [5], the nonlinear up-conversion luminescence in the aqueous suspensions of carbon dots with polyfunctional and carboxylated surfaces synthesized by a hydrothermal method was observed. It was shown that the discovered nonlinear up-conversion luminescence in these carbon dots was caused by two-photon absorption. The optimal laser excitation wavelength at which nonlinear up-conversion luminescence manifests itself most strongly was found. The results obtained can be used to determine the pH of liquid media in a wide range of pH values.

A 40 nm thick film of topological insulator ($Bi_{1.9}Sb_{0.1}Te_2Se$) was used for efficient generation of terahertz radiation under femtosecond laser radiation at 1560 nm excitation in [6]. It was shown that the generation of terahertz radiation in this film occurred due to photo-excited bulk carriers and their relaxation through the surface states in the presence of the constant electric field. Thus, it has been demonstrated that low-cost $Bi_{1.9}Sb_{0.1}Te_2Se$ ultrathin topological insulators, which can be easily integrated into various nanophotonic devices, enable the conversion of femtosecond infrared radiation to terahertz radiation. The research presented in [7] focuses on the development and investigation of a new type of robust photonic topological insulators for laser array applications. It was constructed from identical prism resonators connected to each other. The author experimentally demonstrated a topologically protected propagating state due to the disconnected faces of the edge resonators.

A new and accurate, fully explicit finite-difference time-domain method for modeling a nonlinear electromagnetic medium has been proposed in theoretical work [8]. The developed algorithms describe the interaction of the laser pulse with metals and interfaces of nonlinear dielectric media well, where Kerr and Raman effects, as well as multiphoton ionization and metal dispersion, occur simultaneously. The author also modeled the nonlinear propagation of an ultrafast laser pulse through a dielectric medium using this method.

In [9], the generation of nanosecond photocurrent pulses due to the surface photogalvanic effect (the second-order nonlinear optical phenomenon) was studied in thin semitransparent CuSe/Se films as a function of the angle of incidence and polarization of exciting femtosecond laser pulses at 795 nm. It was found that the evolution of the temporal profile of the helicity-sensitive transverse photocurrent with a change in incident angle strongly depends on polarization. In particular, at circular polarization, the generation of unipolar and bipolar pulses is possible, with the waveforms strongly depending both on the angle of incidence and the sign of circular polarization. A kinetic theory of the edge photogalvanic effect for the intraband electron transport in two-dimensional materials was developed in [10]. It should be noted that a photogalvanic effect in 2D structures can be considered as a low-dimensional analog of the surface photogalvanic effect. The authors obtained an analytical expression for the edge current valid for arbitrary dispersion law and scattering mechanism and analyzed the result for single-layer and bilayer graphene for electron scattering by short-range defects and Coulomb impurities. The authors believe that the edge photogalvanic effect will find applications in fast terahertz radiation and terahertz radiation polarization sensors.

Funding: This work was supported by the Ministry of Education and Science of the Russian Federation (state registration number 1021032422167-7-1.3.2.).

Acknowledgments: I consider it my duty to express my gratitude to all the authors of the articles who sent their insightful papers to this Special Issue. Special thanks to all the reviewers who provided substantive comments and suggestions to improve the quality of the submitted articles. I am grateful to the entire editorial team of *Applied Sciences* for the high editorial quality. I would also like to thank A.E. Fateev for discussing the articles published in this Special Issue.

Conflicts of Interest: The author declares no conflict of interest.

References

1. Ganeev, R.A.; Kuroda, H. Reexamining different factors of the resonance-enhanced high-order harmonic generation in atomic and nanoparticle laser-induced tin plasmas. *Appl. Sci.* **2021**, *11*, 21193. [CrossRef]
2. Ganeev, R.A.; Kim, V.V.; Rao, K.S.; Guo, C. Probing laser plasma dynamics using high-order harmonics generation in carbon-containing nanomaterials. *Appl. Sci.* **2021**, *11*, 2143. [CrossRef]
3. Maydykovskiy, A.; Temiryazeva, M.; Temiryazev, A.; Murzina, T. Nonlinear optical microscopy of interface layers of epitaxial garnet films. *Appl. Sci.* **2023**, *13*, 8828. [CrossRef]
4. Mikheev, G.M.; Vanyukov, V.V.; Mogileva, T.N.; Mikheev, K.G.; Aleksandrovich, A.N.; Nunn, N.A.; Shenderova, O.A. Femtosecond optical nonlinearity of nanodiamond suspensions. *Appl. Sci.* **2021**, *11*, 5455. [CrossRef]
5. Laptinskiy, K.; Khmeleva, M.; Vervald, A.; Burikov, S.; Dolenko, T. Carbon dots with up-conversion luminescence as pH nanosensor. *Appl. Sci.* **2022**, *12*, 12006. [CrossRef]
6. Kuznetsov, K.A.; Safronenkov, D.A.; Kuznetsov, P.I.; Kitaeva, G.K. Terahertz photoconductive antenna based on a topological insulator nanofilm. *Appl. Sci.* **2021**, *11*, 5580. [CrossRef]
7. Kim, P.N.; Fedchenko, D.P.; Rudakova, N.V.; Timofeev, I.V. Tiling photonic topological insulator for laser applications. *Appl. Sci.* **2023**, *13*, 4004. [CrossRef]
8. Moreno, E.; Nguyen, H.D.; Stoian, R.; Colombier, J.P. Full explicit numerical modeling in time-domain for nonlinear electromagnetics simulations in ultrafast laser nanostructuring. *Appl. Sci.* **2021**, *11*, 7429. [CrossRef]
9. Fateev, A.E.; Mogileva, T.N.; Kogai, V.Y.; Mikheev, K.G.; Mikheev, G.M. Incident angle dependence of the waveform of the polarization-sensitive photoresponse in CuSe/Se thin film. *Appl. Sci.* **2022**, *12*, 6869. [CrossRef]
10. Durnev, M.V.; Tarasenko, S.A. Edge currents induced by AC electric Ffeld in two-dimensional dirac structures. *Appl. Sci.* **2023**, *13*, 4080. [CrossRef]

applied sciences

MDPI

Article

Reexamining Different Factors of the Resonance-Enhanced High-Order Harmonic Generation in Atomic and Nanoparticle Laser-Induced Tin Plasmas

Rashid A. Ganeev [1,2,3,4,*] **and Hiroto Kuroda** [1,5,6]

[1] Ophthalmology and Advanced Laser Medical Center, Saitama Medical University, Saitama 350-0495, Japan; kuroda-alt@theia.ocn.ne.jp
[2] Moscow Institute of Physics and Technology, 141701 Dolgoprudny, Russia
[3] Institute of Astronomy, University of Latvia, LV-1586 Riga, Latvia
[4] Department of Physics, Voronezh State University, 394006 Voronezh, Russia
[5] Advanced Laser Technology Inc., Tama-city, Tokyo 206-0014, Japan
[6] Plasma Investigations Group, Aichi Medical University, Nagakute 480-1195, Japan
* Correspondence: rashid_ganeev@mail.ru

Abstract: We reexamine the resonance enhancement of a single harmonic emission during the propagation of ultrafast pulses through atomic and nanoparticle tin-containing laser-induced plasma (LIP). We compare the single atomic Sn and Sn nanoparticle plasmas to demonstrate a distinction in the enhancement factor of the single harmonic in the case of fixed and tunable near-infrared pulses. The analysis of the dynamics of Sn LIP shows the range of optimal delays between heating and driving pulses (130–180 ns), at which the maximal harmonic yield can be achieved. The enhancements of the 17th and 18th harmonics of 806 nm pulses were analyzed in the case of single-color and two-color pumps of LIP, showing up to a 12-fold enhancement of even harmonics in the two-color pump case. We show the enhancement of a single harmonic in the vicinity of the $4d^{10}5s^25p^2P_{3/2} \rightarrow 4d^95s^25p^2$ transitions of Sn II ions and demonstrate how this process depends on the constituency of the plasma components at different conditions of target ablation. The application of tunable (1280–1440 nm) radiation allows for demonstrating the variations of single harmonic enhancement using a two-color pump of Sn-containing LIP.

Keywords: high-order harmonic generation; nanoparticles; tin plasma; resonance enhancement of harmonics

Citation: Ganeev, R.A.; Kuroda, H. Reexamining Different Factors of the Resonance-Enhanced High-Order Harmonic Generation in Atomic and Nanoparticle Laser-Induced Tin Plasmas. *Appl. Sci.* **2021**, *11*, 2193. https://doi.org/10.3390/app11052193

Academic Editor: Alessandro Belardini

Received: 6 February 2021
Accepted: 26 February 2021
Published: 3 March 2021

1. Introduction

Resonance-related effects play an important role in different fields of laser physics and optics. The interest in these effects is related to the possibility of improving various processes during light–matter interactions. One example of the application of the resonance effects in nonlinear optics is the enhancement of high-order harmonic yield during the interaction of strong laser pulses with gases and laser-induced plasmas (LIPs). The studies of this process during the high-order harmonic generation (HHG) of ultrafast laser pulses have been the topic of different publications over the last two decades [1–18]. The HHG model describing the enhancement of the harmonic coinciding with the transition between the ground and the autoionizing states of ions has been developed in [10]. The third (recombination) step of the three-step scenario of HHG was partitioned into two steps: the capture of a laser-accelerated electron onto an autoionizing state of the parent ion, followed by the radiative relaxation of this state to the ground state with the emission of the harmonic photon. Probably, this scenario of resonance HHG could be applied to ionized large particles as well.

HHG in LIPs has actively been studied using different ablated media (bulk metals, organic materials, multiparticle-contained powders, etc. [19–41]). It was mentioned in previous studies that the enhancement of the resonance harmonics compared with the neighboring ones—in the case, of the nanoparticle-containing plasmas—was notably smaller than in the case of the monomer-containing plasmas [33]. Additionally, different issues require further consideration, once one considers the resonance processes as the suitable method for a single harmonic enhancement. Particularly, tin-containing plasmas, which have earlier been examined as a suitable medium for the demonstration of this effect [42,43], allow for analyzing different processes in various constituents of LIPs. Additionally, the availability of tunable laser sources would allow for determining the influence of the detuning out of resonances on the enhancement factor of a single harmonic.

In this paper, we reexamine the single harmonic resonance enhancement during the propagation of the ultrafast pulses through tin-containing LIP. We demonstrate the enhancement of the 17th and 18th harmonics of 806 nm pulses in the case of different (single-color and two-color) pumps and compare the single atomic Sn and Sn nanoparticle LIPs to demonstrate a distinction in the enhancement factor of a single harmonic in the case of tunable near-infrared (NIR) pulses. The application of tunable (1280–1440 nm) radiation allows for demonstrating the variations of single harmonic enhancement using a two-color pump of Sn-containing plasma.

2. Materials and Methods

Bulk tin (Sn) and Sn nanoparticles (NPs; 70 nm, Sigma-Aldrich, St. Louis, MO, USA), glued onto a glass substrate, were ablated by different laser pulses to form LIP, with further HHG during the propagation of femtosecond laser pulses. The part of uncompressed radiation from a Ti: sapphire laser (central wavelength of 806 nm, pulse duration of 370 ps, pulse energy of 10 mJ, and a pulse repetition rate of 10 Hz) was used as a heating pulse (HP) for the target ablation (Figure 1). We also used a Nd: YAG laser for target ablation, using 1064 nm, 10 ns pulses.

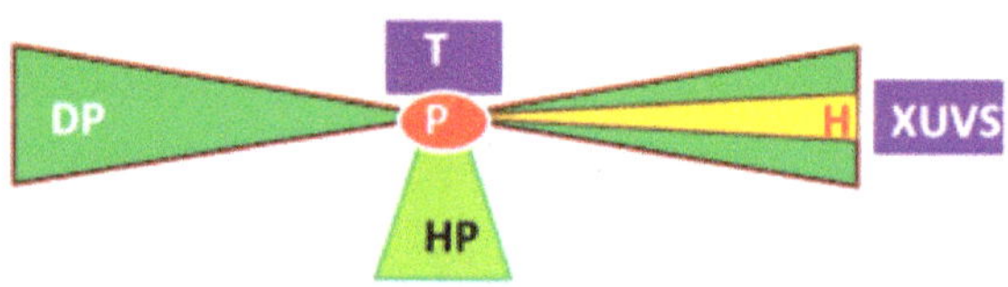

Figure 1. Experimental scheme. DP: driving pulse; HP: heating pulse; T: target; P: laser-induced plasma; H: harmonic emission; XUVS: extreme ultraviolet spectrometer.

The focused compressed (64 fs) driving pulses (DP) were propagated through the LIP at a distance of 0.2 mm above the target surface. Apart from the single-color pump (SCP; 806 nm) of plasmas, we used the two-color pump (TCP; 806 nm and 403 nm) to generate odd and even harmonics. A 0.2 mm-thick beta-barium borate (BBO; type I) crystal was inserted into the vacuum chamber in the path of the focused 806 nm driving pulses to generate a second harmonic (λ = 403 nm, 3% conversion efficiency).

We also used an optical parametric amplifier (OPA), pumped by the above-described laser, to apply the tunable NIR DP for HHG in the LIP. The signal pulses from the OPA were used as the DP (0.8 mJ, 70 fs, 1280 nm). The intensity of the focused NIR DP inside the LIP was 2×10^{14} W cm^{-2}. Most of the experiments using the OPA were carried out using the TCP of the LIP. We used the second harmonic (H2) of the signal pulses to apply the TCP scheme (NIR and H2) for HHG in the plasma.

The harmonic radiation was analyzed using an extreme ultraviolet (XUV) spectrometer containing a gold-coated spherical mirror and a 1200 grooves/mm flat-field grating with variable line spacing. The spectrum was recorded on a micro-channel plate (MCP) detector with a phosphor screen, which was imaged onto a charge-coupled device (CCD) camera. The movement of the MCP along the focusing plane of the flat-field grating allowed for the observation of harmonics in different regions of the XUV spectrometer.

3. Results

3.1. 806 nm Pump

The upper panel of Figure 2 shows the raw image of the spectral distribution of plasma emission during ablation of the tin bulk target, using a 370 ps HP at a fluence of F = 3.2 J cm^{-2}. This spectrum was collected without the propagation of a 64 fs DP through the plasma plume and is shown for the demonstration of various emission lines from highly-ionized tin LIP. The application of such plasma for HHG has two drawbacks: the first disadvantage is associated with strong incoherent radiation in the range of harmonic generation, which can entirely overlap later radiation, and the second disadvantage is related to a deterioration of the optimal phase-matching conditions between driving and harmonic waves. The latter disadvantage leads to a reduction or entire disappearance of harmonic emissions due to a phase mismatch of the interacting waves. In addition, it also affects the conditions for the resonant amplification of harmonics. The propagation of femtosecond pulses through this plasma did not result in harmonic generation or allow for the observation of the weak lower-order harmonics.

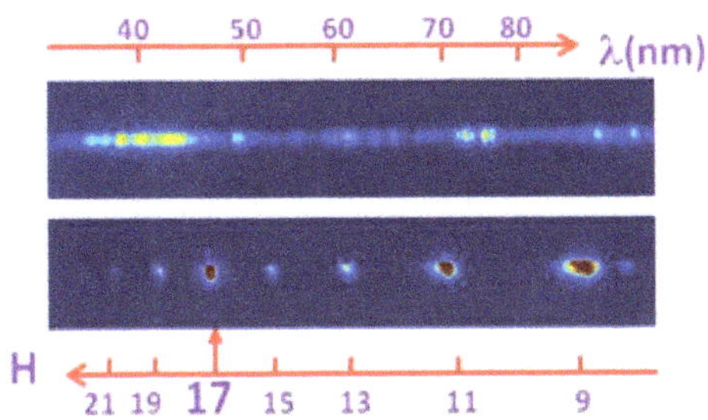

Figure 2. Raw images of plasma (upper panel) and harmonic (bottom panel) emission spectra. Upper and bottom lines show the calibrated wavelength (in nm) and harmonic axes, respectively.

To achieve the efficient generation of harmonics, we used a twice smaller fluence of 370 ps pulses on the target surface (1.6 J cm^{-2}) when no lines attributed to the ionic transitions in this spectral range were observed (bottom panel of Figure 2). The propagation of the DP through such LIP resulted in the generation of an XUV spectrum, comprising the harmonics of 806 nm radiation up to the 21st order (H21). One can see a strong emission attributed to the resonance-enhanced H17 (λ = 47.4 nm), which notably exceeded the nearby harmonics. These studies were performed at a 150 ns delay between the HP and the DP. One can see that the strong H17 almost coincides with the positions of some weak emission lines of the plasma (upper panel of Figure 2). This harmonic was 4 to 10 times stronger than the nearest harmonic orders.

The raw images of the harmonic spectra provide a better comparative view of the plasma and harmonic emissions. The bottom image of Figure 2 shows the common feature of the gradually decaying plateau-like harmonic distribution along the whole studied XUV spectrum except for the single-enhanced harmonic (H17). These raw images were taken at the saturation conditions of the CCD camera for better viewing of the peculiarities of the harmonic distribution. Notice that all line-outs of the harmonic spectra shown in the following figures were taken using the images collected at the unsaturated conditions of registration.

We used two methods to analyze the dynamics of tin plasma spreading out from the target surface. Apart from the plasma emission spectra, we first analyzed the images of the plasma at different moments from the beginning of ablation. Useful information about ablated species can be retrieved by recording the spatial dynamics of the spreading of the plasma plume using a gated intensified charge-coupled device (ICCD) camera, which was used for imaging of the spreading tin LIP within a time interval ranging from 0 to 2000 ns with a variable time gate. In this case, we used 1064 nm, 10 ns pulses as the HP. The used HP energy (E_{HP} = 20 mJ) was higher than that used for the harmonic generation, for better visibility of the plasma plume, while the overall dynamics of the plasma formation and spreading were similar, thus allowing for the determination of the main regularities.

At the initial moments from ablation (i.e., up to 50 ns), the LIP was spatially limited inside the dense and very bright small volume. As it was shown by analysis of the time-resolved spectra of the ns-induced LIP in the visible region [44], a strong continuum emission can be seen at the early periods of plasma dynamics. Neutral emission peaks were not evident at early times in the case of the ns-induced plasma spectra and required longer delays to appear. Additionally, they showed lower intensities compared to the ionic emissions at the early stages. This is attributed to a higher population of the charged ions emitting in the visible region compared to the neutral atoms, though once these ion populations decreased due to recombination processes in the LIP, at later times the emissions from neutral species became a dominating factor. Images in the case of the time delays above 100 ns show a small, expanded laser plume up to a distance of 1.2 mm from the surface of the bulk tin. Due to high brightness, we suppressed emissions by placing neutral filters in front of the ICCD camera. At a period of 200 ns from the ablation, the fastest components of plasma expanded up to a distance up to 2.5 mm. However, their role in harmonic generation is questionable due to a small concentration of these fast components of LIP. We analyzed the images of plasma up to 2000 ns from the beginning of ablation, when a weak, low-dense cloud of plasmas was still seen in the images.

The second method of the plasma dynamics studies was based on the analysis of the variations of harmonic emissions at different delays between the HP and the DP using a two-laser (Ti: sapphire and Nd: YAG) configuration. In this case, HHG spectra from Sn plasma were recorded at different delays between a 10 ns HP and a 64 fs DP. The intensity dependencies of the 11th (H11) and 15th (H15) harmonics on the delay between the HP and the DP are shown in Figure 3A. For this plasma, the maximum delay up to which we were able to observe HHG emission was equal to ~500 ns, at which a very weak emission of the lowest recorded harmonic (H9) was still seen in the XUV spectrum. We observed the clearly determined maximal yields of both H11 and H15 at the same (~150 ns) delay from the beginning of ablation. This pattern can be explained by the presence of the same type of emitters within the whole period of observation, when a decrease of plasma density, due to its expansion, causes the decay of the harmonic yield. The velocity of these emitters, according to optimal delay, can be estimated at ~1.3×10^3 m/s, taking into account the 0.2 mm distance between the target and the axis of DP propagation. Notice that a similar range of the "optimal" delays (140–200 ns) between the HP and the DP corresponded to the observation of the maximal harmonic yield in the case of the ablated Sn NP LIP.

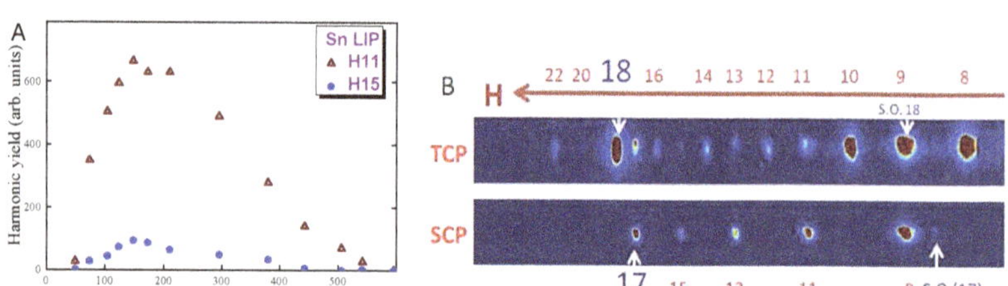

Figure 3. (**A**) Variations of the 11th (open triangles) and 15th (filled circles) harmonic yields at different delays between the heating and driving pulses. The fluence of heating pulses on the tin surface was 1.5 J cm^{-2}. (**B**) Raw images of harmonics in the case of a two-color pump (TCP; 806 nm and 403 nm, upper panel) and a single-color pump (SCP; 806 nm, bottom panel) of tin plasma. Arrows show the maximally enhanced harmonics and their second-order (S.O.) diffractions from the flat-field grating in the case of the two types of pumps.

One can note that during the ablation of bulk Sn, the plasma may contain tin atoms and ions, as well as NPs synthesized during laser ablation. The appearance of Sn nanoparticles was commonly registered at the relatively high fluencies of HP (F $\approx$ 3 J cm^{-2}). The transmission electron microscope (TEM) analysis of the debris deposited on the nearby glass substrates while using this fluence of either picosecond or nanosecond HP revealed the presence of sparsely distributed NPs with sizes ranging between 2 and 10 nm. Once again, we emphasize that this mode of plasma formation was unsuitable for generating harmonics in Sn LIP. The use of the fluence F = 1.5 J cm^{-2}, at which the strongest yield of harmonics from this plasma was achieved, did not reveal the presence of NPs in the deposited debris. Thus, the harmonics, in this case, originated from the atoms and ions of tin.

In these experiments, the femtosecond pulses were delayed with respect to those from the Nd: YAG laser to propagate through the formed plasma at a maximal density of the ejected particles. The synchronization of the femtosecond and nanosecond lasers was accomplished using a digital delay generator, and the delay between the HP and the DP was varied in the range of 0 to 1000 ns. We were also able to analyze the plasma and harmonic dynamics at high delays, spanning from a few units to several hundred microseconds. However, apart from the delay region, in which strong harmonics were observed (100–300 ns, Figure 3A), no harmonics appeared in the microsecond-long delays, thus pointing out the absence of the influence of large mass species on the whole yield of harmonic emissions from the tin LIP.

The application of the TCP to this plasma led to the appearance of a rather stronger difference between the whole set of harmonics and one of them, which was attributed to the influence of a resonance-induced mechanism of the enhancement of a single harmonic in the vicinity of the strong transition of tin ions. In the case of the 0.2 mm-thick BBO, some temporal delay between 806 and 403 nm pulses in the LIP still allowed for the effective interaction, leading to the odd and even harmonics generation, since the trailing part of the second harmonic wave overlaps with the leading part of the 806 nm wave (Figure 3B). Another spectrum appeared in the case of the installation of 0.5 mm-thick BBO on the path of the DP. The harmonic distribution became notably heterogeneous due to an insufficient overlap between the 806 nm and 403 nm pulses, delayed from each other by 95 fs in the plasma area, resulting in the generation of 2(2n + 1) harmonics (H10, H14, H18, and H22), while other even harmonics, corresponding to 4(n + 1) orders (H12, H16, and H20), were almost suppressed.

The most important finding in these TCP studies was the appearance of an exceptionally strong 18th harmonic (upper panel of Figure 3B) notably exceeding the above-analyzed 17th harmonic achieved during the SCP of the Sn LIP. Moreover, second-order diffraction of this harmonic caused a stronger image of H9 (upper right arrow), leading to the impression that H9 became stronger in the case of the TCP compared with the SCP (bottom panel of Figure 3B). The prevalence of H18 over H17 could have been caused by a stronger involvement of the resonance processes in the enhancement of the former harmonic emission. The wavelength of this harmonic (H18, λ = 44.8 nm, E = 27.68 eV) was closer to some $4d^{10}5s^25p^2P_{3/2}\rightarrow4d^95s^25p^2$ ionic transitions of tin, possessing strong oscillator strengths (gf). In the meantime, the gf of one of those transitions [$4d^{10}5s^25p^2P_{3/2}\rightarrow4d^95s^25p^2$ (^{1}D)($^2D_{5/2}$)], at the wavelength of 47.20 nm, has been calculated to be 1.52, and this value is five times larger than other transitions from the ground state of Sn II [45]. From this point of view, one can expect an almost perfect coincidence of an H17 (λ = 47.4 nm) and $4d^{10}5s^25p^2P_{3/2}\rightarrow4d^95s^25p^2$ (^{1}D)($^2D_{5/2}$) transition, and stronger enhancement compared with H18. However, our experiment showed the reverse case when H18 is much stronger than H17.

A few reasons could be responsible for this contradiction. The presence of an even stronger ionic transition of Sn II in the 44 nm range of the XUV spectrum can cause a larger enhancement of H18 with respect to H17. Additionally, the Stark shift of transitions can modify their influence on the harmonic yield. Another option could be the involvement of the Sn III ions [43]. Some of them lie in the vicinity of H18 and their oscillator strengths (gf~0.5) are high enough to cause the resonance enhancement of nearby harmonics. Finally, H18 belongs to the group of 2(2n + 1) even harmonics, which can demonstrate larger conversion efficiency compared with the odd harmonics arising from the 806 nm pump.

Notice that the efficiency of harmonics produced by shorter wavelength sources becomes higher due to the strong wavelength-dependent harmonic yield ($I_{harm} \propto \lambda^{-5}$) ($I_{harm}$ is the harmonic intensity and λ is the driving field wavelength [46]).

3.2. Tunable NIR Pump

One can distinguish the importance of tuning the wavelength of the driving pulses to determine the variations of harmonic yields in the case of resonance-related processes. Therefore, the application of OPA-based tunable sources offers large perspectives in these studies. Below, we apply the tunable NIR pulses from the OPA and their second harmonics for the analysis of resonance enhancement in Sn LIP, as well as in Sn NP LIP.

High-order harmonic generation in two media using TCP pulses was optimized by different means (i.e., the fluence of HP, the intensity of DP, the delay between the HP and the DP, the application of BBO crystals of different thickness, etc.). Particularly, relatively stronger ablation of the Sn NP target allowed for the observation of the enhancement of odd and two neighboring harmonics in the vicinity of a strong ionic transition of tin, contrary to the case of the weak ablation of this nanoparticle-containing target. Our TEM studies of Sn NP debris deposited on nearby glass slides showed the presence of the small NPs ranging from 4 to 14 nm. These NPs appeared in the TEM images at the optimal fluencies of HP ($F = 1.4$–1.7 J cm^{-2}), when the maximal harmonic yield was achieved.

The harmonics in the case of the NIR pulses (1 mJ, 1300 nm) were significantly weaker than in the case of the 3 mJ, 806 nm pump, due to a higher intensity of DP in the latter case, and the $I_{harm} \propto \lambda^{-5}$ rule. The comparison of these two pumps at similar energies of pulses (1 mJ) showed a four-fold decrease of harmonic yield from the plasma produced on the bulk Sn target in the case of 1320 nm pulses compared with the 806 nm pulses, which, to some extent, corresponds to the theoretical prediction of this ratio (11), taking into account a difference in the spatial distributions of two focused beams in the plasma area. Meanwhile, the harmonic cutoff in the case of the NIR pulses was larger than in the case of 806 nm pulses, in accordance with the expectations in the extension of this parameter for the longer-wavelength pump ($E_{cutoff} \propto \lambda^2$). The harmonic cutoffs in the case of the 806 nm and NIR pump (~1300 nm) pulses were in the range of 42 nm (E = 29.5 eV) and 38 nm (E = 32.6 eV), respectively, while the expected cutoff was calculated to be 76 eV. However, the very small conversion efficiency of higher-order harmonics in the case of NIR pulses did not allow for the observation of harmonics well below the 40 nm spectral region. Thus, the use of 1-mJ NIR pulses for plasma harmonic studies caused the generation of weak odd harmonics in the longer-wavelength range of the XUV spectrum. For this reason, we used the TCP (NIR and H2) of plasma, which proved to enhance the odd harmonic yield. The extension of the harmonic cutoff compared with the single-color pump, the ten-fold growth of the yield of odd harmonics, the comparable harmonic intensities for the odd and even orders along the whole range of generation in spite of using the 0.5 mm-long crystal, contrary to the case of the 806 nm pump, and the tuning of harmonics that allowed for the optimization of resonance-induced single harmonic generation were among the attractive features of these experiments.

The raw images of the harmonic spectra presented in Figure 4 were obtained using NIR pulses (λ = 1370 nm) and the second harmonic of this radiation (λ = 685 nm) during their propagation through the plasma produced on the bulk tin and tin NP targets. As in the case of the 806 nm pump of the Sn LIP, we observed the enhanced harmonic in the region of 47 nm (Figure 4, upper panel). At the used pump conditions, it was the 29th harmonic (H29) of 1370 nm radiation. The application of ablated tin nanoparticles as the media for harmonic generation at the fluence similar to the one used for the ignition of plasma in the case of the tin bulk target (F = 1.4 J cm^{-2}) allowed for the increase of the conversion efficiency of a whole set of harmonics (Figure 4, middle panel). The harmonics were extended up to the 31st order. However, no variations from the featureless plateau-like pattern of the envelope of the gradually decaying harmonic distribution were achieved in this case, except for some insignificant differences in the intensities of H29 and neighboring orders. This difference in the harmonic spectra produced from the plasmas ignited on the bulk Sn and Sn NP targets was attributed to an insufficient amount of ionized single-particle species in the latter case.

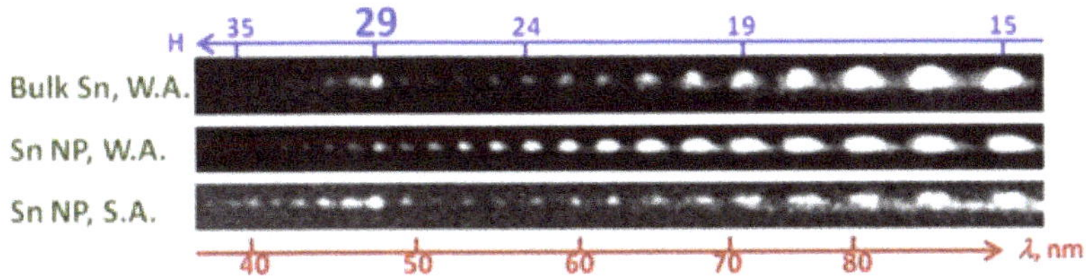

Figure 4. Raw images of harmonic spectra in the case of 1370 nm and 685 nm pumps of tin-containing plasmas. Upper panel: Plasma produced on the tin bulk target at the fluence F = 1.4 J cm^{-2}. Middle panel: Plasma produced on the Sn nanoparticle (NP) target at the fluence F = 1.4 J cm^{-2}. Bottom panel: Plasma produced on the Sn NP target at the fluence F = 2 J cm^{-2}. "W.A." and "S.A." correspond to weak and strong ablation of the Sn-containing target.

Changing the ablation conditions of the Sn NP target can lead to the appearance of single-particle species and additional excited states of ions that can, to some extent, amplify the neighboring harmonics. Among different types of ablated substances, tin plasma can be considered as an interesting subject for study since it allows for the modification of the excitation of various autoionizing states, depending on the ablation conditions of the bulk target. The variations of target ablation conditions could allow for the analysis of the nonlinear response of such plasma by comparing the amplification of different harmonics, provided that they coincide or remain close to those autoionizing states responsible for harmonic enhancement. It is interesting to consider this process in the case of the ablation of the targets containing nanoparticles since the modification of their morphology or even the disintegration of NPs onto the small species and atoms can drastically change the distribution of harmonics in the region of interest, where strong ionic transitions can cause the enhancement of harmonics in the case of an atomic/ionic plasma medium.

The fluence of heating 370 ps pulses was increased up to F = 2 J cm^{-2} to produce a sufficient amount of Sn ions in the plasma ignited on the NP-containing surface. These conditions of ablation proved to be non-optimal for the efficient generation of harmonics. In this case, we observed the worsening of the phase-matching conditions, leading to a decrease in the conversion efficiency of the lower-order harmonics (bottom panel of Figure 4). The appearance of a large number of free electrons at stronger ablation caused the phase mismatch between the harmonic and driving waves, due to the growing group of velocity dispersion in such LIPs. In the meantime, we achieved the enhancement of a single harmonic and a few following harmonic orders in the area of strong ionic transitions of tin. H29, in this case, was notably stronger than any nearby lower-order harmonic, down to H19. Probably, at these conditions of nanoparticle ablation, they disintegrated in smaller nanoparticles, clusters, and single-atomic species, which can be presented in an ionic state, thus allowing the resonance-related enhancement of harmonics.

To further analyze the observed process of the variation of harmonic distribution at different conditions of DP–LIP interactions, we performed the tuning of the driving NIR pulses in the case of two plasmas. The line-outs of Figure 5A demonstrate the harmonic spectra generated in the plasmas produced on the bulk tin and tin NP powder. The three upper panels show the HHG spectra in the plasma produced on the tin bulk target at optimal conditions of ablation when the maximal yield of harmonics was achieved. The upper panel was obtained in the case of the SCP (1355 nm) of the plasma produced on the bulk Sn. One can see the weakly enhanced harmonic (H29) lying far from the whole set of the observed lower-order harmonics. We also used the TCP (1355 nm and 678 nm) of plasma at a similar wavelength of NIR pulses. Again, as in the case of the SCP, H29 and two neighboring harmonics can be clearly distinguished in the 46–48 nm spectral range. The tuning of the NIR pulses towards the shorter wavelength region (1335 nm) caused the change in the maximally enhanced harmonic order in the case of the TCP (H28, 1335 nm and 668 nm pumps, third panel) and allowed for the generation of a higher yield of this harmonic compared with the neighboring ones.

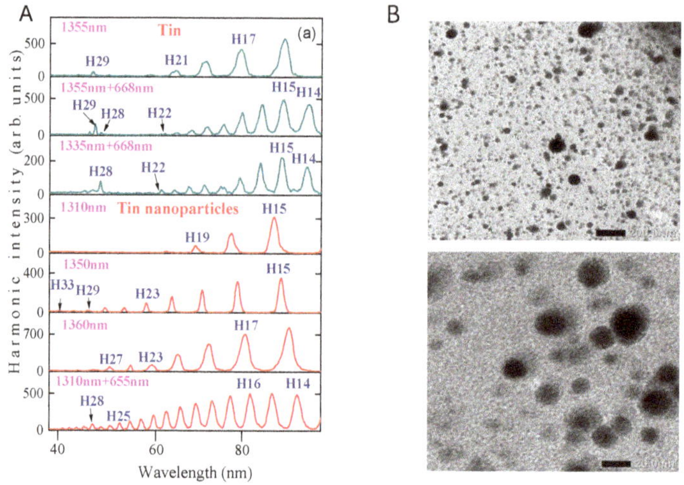

Figure 5. (**A**) Line-outs of harmonic spectra generated in the plasmas produced on the bulk tin and tin NP powder using tunable near-infrared (NIR) driving pulses. First panel from the top: SCP (1355 nm) of the plasma produced on the bulk Sn. Second panel: TCP (1355 nm and 678 nm) of the plasma produced on the bulk Sn. Third panel: TCP (1335 nm and 668 nm) of the plasma produced on the bulk Sn. Fourth panel: SCP (1320 nm) of the plasma produced on the tin NP powder. Fifth panel: SCP (1350 nm) of the plasma produced on the tin NP powder. Sixth panel: SCP (1370 nm) of the plasma produced on the tin NP powder. Seventh panel: TCP (1320 nm and 660 nm) of the plasma produced on the tin NP powder. (**B**) TEM images of deposited tin NPs during ablation of nanoparticle powder at an optimal fluence of HP. The black markers correspond to 200 nm (upper panel) and 20 nm (bottom panel).

Similar studies were performed in the case of Sn NP LIP. The application of nanoparticles was a part of our efforts to demonstrate the fact that the presence of tin in plasma does not necessarily guarantee the enhancement of a single harmonic. The decrease of oscillator strength of ionic transitions, as well as the detuning of transitions, can cause the difference in the nonlinear optical response of this medium. Meanwhile, the expectation in the application of NPs was related to the stronger effect of single harmonic enhancement, while our studies showed the reverse effect.

The distribution of the harmonic spectrum was notably changed as soon as we ablated the Sn NP target. One can assume similar conditions for the resonance enhancement of H29 in this aggregated structure. However, in the case of the SCP of Sn NP LIP, we did not observe this process. At the used fluence of heating pulses (1.5 J cm^{-2}), we did not see a strengthening of this harmonic, which points out the insignificant amount of the ionic tin component during the disintegration of Sn NPs in these conditions of ablation, independently of the wavelength of NIR DP.

Specifically, the experimental observations were as follows. The fourth panel from the top of Figure 5A shows the harmonic line-out in the case of the SCP (1320 nm) of the plasma produced on the tin NP powder. Only a few harmonics up to H19 generated at these conditions in the studied spectral range (38–89 nm). The optimization of plasma formation and stronger NIR pulses tuned towards 1350 nm allowed for achieving the harmonics up to H33 (fifth panel). At these conditions, no enhancement of the harmonic lying close to the 47 nm region was achieved. The tunings of the DP (1370 nm) also did not result in the generation of enhanced harmonics in the 47 nm region (sixth panel). Finally, the seventh panel shows HHG using the TCP (1320 nm and 660 nm) of the plasma produced on the tin NP powder. Again, we did not see the specific feature of single harmonic enhancement in the above-mentioned XUV range, but rather observed the gradual decay of harmonic yield down to H32. An insignificantly larger yield of H28 shown in this panel may indicate the involvement of the resonance process in the generation of this harmonic (similar to the raw image shown in the middle panel of Figure 4 in the case of 1370 nm and 685 nm pumps). The appearance of all even harmonics, in the case of NIR pulses, using thicker (0.5 mm) BBO crystal, revealed a sufficient spatiotemporal overlap of tunable 1300–1370 nm radiation and its second harmonic emission in LIPs due to a lesser difference in the group velocity dispersion of these pulses in the NIR range.

At these conditions of Sn NP ablation, the analysis of debris from the ablated area showed the presence of notably smaller NPs compared with the original 70 nm species glued on the glass substrates. Figure 5B shows two TEM images with different magnification factors of the debris deposited on the nearby substrates, which allow for the determination of the presence of a broad range of the dimensions of nanoparticles (between a few units of nm up to 15 nm). Some smaller clusters yet resolved in these images, being in a plasma state, could be responsible for the resonance enhancement of a single harmonic.

The studies of harmonics using idler pulses (1600–2400 nm) were carried out as well. The harmonics were generated using the wavelengths of a pump close to the regenerative regime of the OPA (1600 nm), due to higher pulse energy. The experiments were carried out in the 1600–1900 nm range. The application of these longer wavelength pumps led to a notable decrease in harmonic generation efficiency due to the $I_{harm} \propto \lambda^{-5}$ rule. Only the two-color pump configuration was suitable, in this case, to generate odd and even harmonics, while the application of the single-color pump led to a notably weaker harmonic emission. This observation, as well as the above-presented data, points out the attractiveness of the two-color pump over the single-color pump for harmonic generation in laser-induced plasmas, independently on the spectral range of emissions of the fundamental radiation.

4. Discussion

The most important feature of these studies was the availability in the analysis of the excited states influencing the harmonic distribution at different excitation conditions of the ablating target. We presented the results of the harmonic spectra modification using tunable driving two-color (NIR and H2) orthogonally polarized pulses, which led to the observation of the resonance enhancement of some harmonic orders caused by different states of plasmas. Studying resonance-induced enhancement of harmonics using tunable NIR and H2 pulses provides the opportunity to fine-tune this high-order nonlinear optical process for wavelength-dependent enhancement of the harmonic yield. This advantage underlines the attractiveness of the analysis of metals, semiconductors, multi-particle systems, and so forth, and, particularly, of the oscillator strengths of some ionic transitions using the plasma HHG technique. Thus, this approach could be considered as a new method of nonlinear spectroscopy.

The second field (403 nm or 685 nm) breaks the inversion symmetry, which allows for the observation of even harmonics generation forbidden in the case of the 806 nm and 1370 nm pumps. The interaction of a strong fundamental wave (806 nm or 1370 nm) and a weak second harmonic wave (403 nm or 685 nm) in the LIP becomes a sufficient factor to drastically change the spectral pattern of harmonic emissions. Notice that, in this case, the yields of even harmonics were comparable with those of odd ones, in spite of a ~30:1 ratio of the energies of two pumps. The demonstrated effect shows that aweak second field serves as a trigger for the notable change of this nonlinear optical process.

Most of our studies were performed at the fluence F = 1.6 J cm^{-2}. As we mentioned during the description of the results shown in Figure 4, the growth of the fluence of the 370 ps heating pulses on the target containing Sn NPs allowed for achieving the conditions when the resonance-induced enhancement of the harmonic in the vicinity of strong ionic transitions of Sn becomes pronounced. The disintegration of large NPs into smaller ingredients (atoms/ions, clusters, and tiny NPs) at these conditions of ablation resulted in the involvement of those ionized species in the process of HHG, followed by the observation of the enhancement of some specific harmonic (either H28 or H29) of the used tunable NIR radiation. The significant suppression of the "resonant"-induced harmonic gain (either the 17th order of the 806 nm pump or 28th and 29th orders of the NIR pump) could have been caused by the shift of the above-discussed transition. However, our experiments using tunable NIR pulses at the fluence of the HP at F = 1.6 J cm^{-2} did not reveal the resonance-related processes lead to the enhancement of the harmonic yield. This observation suggests that the tuning of the ionic transition should not be the main factor in leading to the cancellation of the resonance-related effect.

Another factor may be a decrease in the oscillator strength of this transition. The transitions of Sn II and Sn III can be either detuned or weakened once the original species become presented in the form of nanoparticles. The reduction mechanisms of harmonic amplification can be related to either the reduction of the oscillator strength of this transition or the detuning from the wavelength of "resonant" harmonics, or both.

The macroscopic dispersion properties of the plasma could be another key factor for the formation of the resonance-induced enhancement of a single harmonic in tin plasma, as was suggested in [43]. Recent studies of two-component plasmas allowed for analyzing this opportunity in the case of one of the components (Sn) allowing for the generation of enhanced harmonics and the other component (Pb) causing the modification of plasma dispersion properties [47]. The observation of the enhancement of the even (H18) harmonic and the odd (H17) harmonic, in the case of the TCP and SCP of two-component plasma, have demonstrated that the macroscopic properties of the plasma cannot be considered as the origin of the single harmonic enhancement. In other words, the co-existence of lead plasma does not destroy the resonance effect and single harmonic enhancement caused by tin, which suggests the diminished role of the macroscopic effect in the modification of the resonance-enhanced properties of harmonics in the latter component of plasma.

We did not analyze the size-dependent effect of the harmonic yield in the case of nanoparticle-containing plasmas. Meanwhile, the assumption of the influence of a strong local field in the case of smaller-sized nanoparticles, like quantum dots, can predict the growth of harmonic yields in the latter case compared with the large nanoparticles. This assumption can be supported by the vision of HHG as a three-step process, when only atoms on the surface of nanoparticles efficiently participate in harmonic generation, while inner atoms cannot be considered as the emitters of harmonics due to their strong absorbance of generated radiation. The most probable scenario in determining the optimal sizes of atomic aggregates for HHG is the demonstration of the attractiveness of the species with sizes of 1–5 nm, when the quantum confinement effect allows for the increase of the local field-related nonlinearity while limiting the negative effects attributed to the large (20–100 nm) nanoparticles. This assumption has recently been confirmed during the application of metal sulfide quantum dots for HHG [48,49].

There are a few studies of HHG in the plasmas produced on the nanoparticle-coated targets. The aim of those studies was to elevate the nanoparticle cloud by laser ablation (like in our experiments) and to analyze the nonlinear conversion of infrared pulses towards the extreme ultraviolet range. Meanwhile, the most intriguing method would be the propagation of strong laser pulses through nanoparticle-coated thin films. Interest in HHG in solids has been increased dramatically, though this process was demonstrated using the mid-infrared laser sources. The difficulties related to the absorption of generated harmonics in UV and XUV ranges during propagation through solids can be diminished while using the thin films containing nanoparticles.

The main difference of the present study compared to refs 42 and 43 is the analysis of resonance enhancement of the harmonics in the region of ~47 nm during the propagation of the femtosecond pulses through the tin plasma. The analysis, particularly, includes the application of the tunable source in the near-infrared range, which allowed for the tuning of the order of the harmonic being enhanced (from the 28th to the 29th order), which allowed for the identification of the spectral range where the highest enhancement of single harmonics was achieved. The important finding in the case of the two-color pump study was the appearance of the exceptionally strong 18th harmonic of 806 nm radiation notably exceeding the earlier analyzed 17th harmonic achieved during the single-color pump of Sn laser-induced plasma. Our present study shows that the closeness of the wavelength of the harmonic with the ionic transition possessing large oscillator strength (as in the case of the almost perfect coincidence of the H17 (λ = 47.4 nm) and $4d^{10}5s^25p^2P_{3/2}{\rightarrow}4d^95s^25p^2$ $(^1D)(^2D_{5/2})$ transition (λ = 47.8 nm)) does not necessarily lead to the stronger enhancement compared with the neighbor harmonic (H18, in our case). Notice that the latter harmonic became available due to the interaction of the weak second field (403 nm) with the strong fundamental wave (806 nm).

This contradiction was considered in our paper and different scenarios were suggested, which can, to some extent, clarify the observation of the large enhancement of the higher-order harmonic (H18) compared with the one coinciding with the strong ionic transition. The decrease of the oscillator strength of this transition, in the case of nanoparticle and molecular states of tin, as well as the influence of the collective processes when the harmonic in the region of anomalous dispersion of the ionic transition becomes quasi-phase-matched with the driving field, can be the reasons of the observed contradiction with the earlier accepted vision of the resonance-enhancement of a single harmonic in plasma. Thus, the meaning of the reexamination of the earlier reported resonance enhancement of a single harmonic is related to the approach differing from earlier suggested theories and assumptions of the necessity of perfect coincidence of the harmonic and transition wavelengths [9,10,28,33,38,42,43].

As it was underlined, the application of a near-infrared single-color pump did not result in efficient tunable odd harmonics generation to demonstrate the effect of the tunable single harmonic enhancement, due to the reasons described in our manuscript. Because of this, we used only a two-color pump (NIR and H2) to extend the harmonic emissions up to the spectral region where the enhancement of harmonics can be achieved.

The important novelty in this approach is the application of the tunable source of femtosecond pulses, allowing for the analysis of the most suitable conditions for single harmonic enhancement. The comparison of HHG in atomic and nanoparticle plasmas is another approach, allowing for the demonstration of the role of the morphology of harmonic emitters on the enhancement factor of a single harmonic.

5. Conclusions

In conclusion, we have compared the resonance-induced enhancement of single harmonic emissions during the propagation of ultrafast pulses through tin-containing laser-induced plasmas. The analysis of the dynamics of LIPs produced on a Sn bulk target and a Sn NP target has shown the range of the optimal delays between heating and driving pulses (130–180 ns), at which the maximal harmonic yield can be achieved. The similarity of this parameter for two plasmas allows us to suggest the mechanism of energy transfer during ablation, resulting in similar kinetic energy of each ingredient of ejected species.

The enhancement of the 17th and 18th harmonics of 806 nm pulses was analyzed in the case of single-color and two-color pumps of plasma, showing up to a 12-fold enhancement of even harmonics compared with the neighboring orders in the latter case. We have compared the single-atomic Sn and Sn nanoparticles-containing plasmas to demonstrate a distinction in the enhancement factor of the single harmonic in the case of tunable near-infrared pulses. We have shown the enhancement of a single harmonic in the vicinity of the $4d^{10}5s^25p^2P_{3/2}{\rightarrow}4d^95s^25p^2$ transitions of Sn II ions and demonstrated how this process depends on the constituency of the plasma components at different conditions of the target ablation.

The application of tunable (1280–1440 nm) radiation from the optical parametric amplifier allowed us to demonstrate the variations of single harmonic enhancement using a tunable two-color pump of Sn-containing plasma. Our studies using tunable NIR pulses at the fluence of an HP of F = 1.6 J cm^{-2} did not reveal that resonance-related processes lead to the enhancement of the harmonic yield, which suggests that the tuning of ionic transitions should not be the main factor leading to the cancellation of the resonance-related effect. Meanwhile, the presence of tiny NPs and clusters and single-particle species appeared in LIPs during the disintegration of large (70 nm) original aggregates of Sn atoms allows for the formation of conditions for the involvement of tin ions in the enhancement of single harmonics in the vicinity of ionic transitions possessing strong oscillator strength.

Author Contributions: Conceptualization, R.A.G., H.K.; methodology, R.A.G.; formal analysis, R.A.G., H.K.; investigation, R.A.G., H.K.; writing—original draft preparation, R.A.G.; writing—review and editing, R.A.G., H.K.; visualization, R.A.G.; supervision, H.K.; funding acquisition, R.A.G., H.K. All authors have read and agreed to the published version of the manuscript.

Funding: This research was funded by JSPS KAKENHI (grant number 24760048), ERDF (grant number 1.1.1.5/19/A/003), State Assignment to Higher Educational Institutions of Russian Federation (grant number FZGU-2020-0035), grant and the Common Research Facility at the American University of Sharjah (grant number FRG AS1801).

Institutional Review Board Statement: Not applicable.

Informed Consent Statement: Not applicable.

Data Availability Statement: The data that support the findings of this study are available from the corresponding author upon reasonable request.

Conflicts of Interest: The authors declare no conflict of interest.

References

1. Sanpera, A.; Watson, J.B.; Lewenstein, M.; Burnett, K. Harmonic-generation control. *Phys. Rev. A* **1996**, *54*, 4320–4328. [CrossRef]
2. Taïeb, R.; Véniard, V.; Wassaf, J.; Maquet, A. Roles of resonances and recollisions in strong-field atomic phenomena. II. High-order harmonic generation. *Phys. Rev. A* **2003**, *68*, 033403. [CrossRef]
3. Milošević, D.B. Theoretical analysis of high-order harmonic generation from a coherent superposition of states. *J. Opt. Soc. Am. B* **2006**, *23*, 308–315. [CrossRef]
4. Milošević, D.B. High-energy stimulated emission from plasma ablation pumped by resonant high-order harmonic generation. *J. Phys. B* **2007**, *40*, 3367–3376. [CrossRef]
5. Ivanov, I.A.; Kheifets, A.S. Resonant enhancement of generation of harmonics. *Phys. Rev. A* **2008**, *78*, 053406. [CrossRef]
6. Kulagin, I.A.; Usmanov, T. Efficient selection of single high-order harmonic caused by atomic autoionizing state influence. *Opt. Lett.* **2009**, *34*, 2616–2618. [CrossRef]
7. Chen, J.; Zeng, B.; Liu, X.; Cheng, Y.; Xu, Z. Wavelength scaling of high-order harmonic yield from an optically prepared excited state atom. *New. J. Phys.* **2009**, *11*, 113021. [CrossRef]
8. Frolov, M.F.; Manakov, N.L.; Starace, A.F. Potential barrier effects in high-order harmonic generation by transition-metal ions. *Phys. Rev. A* **2010**, *82*, 023424. [CrossRef]
9. Milošević, D.B. Resonant high-order harmonic generation from plasma ablation: Laser intensity dependence of the harmonic intensity and phase. *Phys. Rev. A* **2010**, *81*, 023802. [CrossRef]
10. Strelkov, V. Role of autoionizing state in resonant high-order harmonic generation and attosecond pulse production. *Phys. Rev. Lett.* **2010**, *104*, 123901. [CrossRef]
11. Tudorovskaya, M.; Lein, M. High-order harmonic generation in the presence of a resonance. *Phys. Rev. A* **2011**, *84*, 013430. [CrossRef]
12. Ackermann, P.; Münch, H.; Halfmann, T. Resonantly-enhanced harmonic generation in argon. *Opt. Express* **2012**, *20*, 13824–13832. [CrossRef]
13. Ngoko Djiokap, J.M.; Starace, A.F. Resonant enhancement of the harmonic-generation spectrum of beryllium. *Phys. Rev. A* **2013**, *88*, 053412. [CrossRef]
14. Ganeev, R.A.; Wang, Z.; Lan, P.; Lu, P.; Suzuki, M.; Kuroda, H. Indium plasma in the single- and two-color mid-infrared fields: Enhancement of tunable harmonics. *Phys. Rev. A* **2016**, *93*, 043848. [CrossRef]
15. Wahyutama, I.S.; Sato, T.; Ishikawa, K.L. Time-dependent multiconfiguration self-consistent-field study on resonantly enhanced high-order harmonic generation from transition-metal elements. *Phys. Rev. A* **2019**, *99*, 063420. [CrossRef]
16. Cipura, F.; Halfmann, T. Resonantly enhanced harmonic generation via dressed states with large Autler–Townes splitting. *J. Opt. Soc. Am. B* **2019**, *36*, 2777–2783. [CrossRef]
17. Ngoko Djiokap, J.M.; Starace, A.F. Origin of the multiphoton-regime harmonic-generation plateau structure. *Phys. Rev. A* **2020**, *102*, 013103. [CrossRef]
18. Liang, J.; Lai, Y.H.; Fu, W.; Shan, Y.; Yu, W.; Guo, C. Observation of resonance-enhanced high-order harmonics from direct excitation of metal nanoparticles with femtosecond pulses. *Phys. Rev. A* **2020**, *102*, 053117. [CrossRef]

19. Akiyama, Y.; Midorikawa, K.; Matsunawa, Y.; Nagata, Y.; Obara, M.; Tashiro, H.; Toyoda, K. Generation of high-order harmonic using laser-produced rare-gas-like ions. *Phys. Rev. Lett.* **1992**, *69*, 2176–2779. [CrossRef]
20. Kubodera, S.; Nagata, Y.; Akiyama, Y.; Midorikawa, K.; Obara, M.; Tashiro, H.; Toyoda, K. High-order harmonic generation in laser-produced ions. *Phys. Rev. A* **1993**, *48*, 4576–4581. [CrossRef]
21. Wahlström, C.-G.; Borgström, S.; Larsson, J.; Pettersson, S.-G. High-order harmonic generation in laser-produced ions using a near-infrared laser. *Phys. Rev. A* **1995**, *51*, 585–592. [CrossRef] [PubMed]
22. Theobald, W.; Wülker, C.; Schäfer, F.R.; Chichkov, B.N. High-order harmonic generation in carbon vapor and low charged plasma. *Opt. Commun.* **1995**, *120*, 177–185. [CrossRef]
23. Ganeev, R.A.; Redkorechev, V.I.; Usmanov, T. Optical harmonics generation in low-temperature laser produced plasma. *Opt. Commun.* **1997**, *135*, 251–258. [CrossRef]
24. Krushelnick, K.; Tighe, W.; Suckewer, S. Harmonic generation from ions in underdense aluminum and lithium-fluorine plasmas. *J. Opt. Soc. Am. B* **1997**, *14*, 1687–1695. [CrossRef]
25. Ganeev, R.A.; Singhal, H.; Naik, P.A.; Chakravarty, U.; Arora, V.; Chakera, J.A.; Khan, R.A.; Raghuramaiah, M.; Kumbhare, S.R.; Kushwaha, R.P.; et al. Optimization of the high-order harmonics generated from silver plasma. *Appl. Phys. B* **2007**, *87*, 243–250. [CrossRef]
26. Singhal, H.; Arora, V.; Rao, B.S.; Naik, P.A.; Chakravarty, U.; Khan, R.A.; Gupta, P.D. Dependence of high-order harmonic intensity on the length of preformed plasma plumes. *Phys. Rev. A* **2009**, *79*, 023807. [CrossRef]
27. Pertot, Y.; Elouga Bom, L.B.; Bhardwaj, V.R.; Ozaki, T. Pencil lead plasma for generating multimicrojoule high-order harmonics with a broad spectrum. *Appl. Phys. Lett.* **2011**, *98*, 101104. [CrossRef]
28. Elouga Bom, L.B.; Haessler, S.; Gobert, O.; Perdrix, M.; Lepetit, F.; Hergott, J.-F.; Carré, B.; Ozaki, T.; Salières, P. Attosecond emission from chromium plasma. *Opt. Express* **2011**, *19*, 3677–3682. [CrossRef] [PubMed]
29. Pertot, Y.; Chen, S.; Khan, S.D.; Elouga Bom, L.B.; Ozaki, T.; Chang, Z. Generation of continuum high-order harmonics from carbon plasma using double optical gating. *J. Phys. B* **2012**, *45*, 074017. [CrossRef]
30. Kumar, M.; Singhal, H.; Chakera, J.A.; Naik, P.A.; Khan, R.A.; Gupta, P.D. Study of the spatial coherence of high order harmonic radiation generated from preformed plasma plumes. *J. Appl. Phys.* **2013**, *114*, 033112. [CrossRef]
31. Ganeev, R.A.; Suzuki, M.; Kuroda, H. Quasi-phase-matching of high-order harmonics in multiple plasma jets. *Phys. Rev. A* **2014**, *89*, 033821. [CrossRef]
32. Singhal, H.; Naik, P.A.; Kumar, M.; Chakera, J.A.; Gupta, P.D. Enhanced coherent extreme ultraviolet emission through high order harmonic generation from plasma plumes containing nanoparticles. *J. Appl. Phys.* **2014**, *115*, 033104. [CrossRef]
33. Ganeev, R.A.; Suzuki, M.; Yoneya, S.; Kuroda, H. Resonance-enhanced harmonic generation in nanoparticle-containing plasmas. *J. Phys. B* **2015**, *48*, 165401. [CrossRef]
34. Rosenthal, N.; Marcus, G. Discriminating between the role of phase matching and that of the single-atom response in resonance plasma-plume high-order harmonic generation. *Phys. Rev. Lett.* **2015**, *115*, 133901. [CrossRef]
35. Fareed, M.A.; Thiré, N.; Mondal, S.; Schmidt, B.E.; Légaré, F.; Ozaki, T. Efficient generation of sub-100 eV high-order harmonics from carbon molecules using infrared laser pulses. *Appl. Phys. Lett.* **2016**, *108*, 124104. [CrossRef]
36. Fareed, M.A.; Strelkov, V.V.; Thiré, N.; Mondal, S.; Schmidt, B.E.; Légaré, F.; Ozaki, T. High-order harmonic generation from the dressed autoionizing states. *Nature Commun.* **2017**, *8*, 16061. [CrossRef] [PubMed]
37. Wöstmann, M.; Splitthoff, L.; Zacharias, H. Control of quasi-phase-matching of high-harmonics in a spatially structured plasma. *Opt. Express* **2018**, *26*, 14524–14531. [CrossRef] [PubMed]
38. Fareed, M.A.; Strelkov, V.V.; Singh, M.; Thiré, N.; Mondal, S.; Schmidt, B.E.; Légaré, F.; Ozaki, T. Harmonic generation from neutral manganese atoms in the vicinity of the giant autoionization resonance. *Phys. Rev. Lett.* **2018**, *121*, 023201. [CrossRef]
39. Abdelrahman, Z.; Khohlova, M.A.; Walke, D.J.; Witting, T.; Zair, A.; Strelkov, V.V.; Marangos, J.P.; Tisch, J.W.G. Chirp-control of resonant high-order harmonic generation in indium ablation plumes driven by intense few-cycle laser pulses. *Opt. Express* **2018**, *26*, 15745–15755. [CrossRef]
40. Kumar, M.; Singhal, H.; Chakera, J.A. High order harmonic radiation source for multicolor extreme ultraviolet radiography of carbon plumes. *J. Appl. Phys.* **2019**, *125*, 155902. [CrossRef]
41. Singh, M.; Fareed, M.A.; Laramée, A.; Isgandarov, E.; Ozaki, T. Intense vortex high-order harmonics generated from laser-ablated plume. *Appl. Phys. Lett.* **2019**, *115*, 231105. [CrossRef]
42. Suzuki, M.; Baba, M.; Ganeev, R.; Kuroda, H.; Ozaki, T. Anomalous enhancement of single high-order harmonic using laser ablation tin plume at 47 nm. *Opt. Lett.* **2006**, *31*, 3306–3308. [CrossRef]
43. Ganeev, R.A.; Strelkov, V.V.; Hutchison, C.; Zaïr, A.; Kilbane, D.; Khokhlova, M.A.; Marangos, J.P. Experimental and theoretical studies of two-color pump resonance-induced enhancement of odd and even harmonics from a tin plasma. *Phys. Rev. A* **2012**, *85*, 023832. [CrossRef]
44. Freeman, J.R.; Harilal, S.S.; Diwakar, P.K.; Verhoff, B.; Hassanein, A. Comparison of optical emission from nanosecond and femtosecond laser produced plasma in atmosphere and vacuum conditions. *Spectrochim. Acta B* **2013**, *87*, 43. [CrossRef]
45. Duffy, G.; van Kampen, P.; Dunne, P. 4d-5p transitions in the extreme ultraviolet photoabsorption spectra of SnIIand SnIII. *J. Phys. B* **2001**, *34*, 3171–3182. [CrossRef]
46. Lan, P.; Takahashi, E.; Midorikawa, K. Wavelength scaling of efficient high-order harmonic generation by two-color infrared laser fields. *Phys. Rev. A* **2010**, *81*, 061802. [CrossRef]

47. Liang, J.; Venkatesh, M.; Boltaev, G.S.; Ganeev, R.A.; Lai, Y.H.; Guo, C. Investigation of resonance-enhanced high-order harmonics by two-component laser-produced plasmas. *Atoms* **2021**, *9*, 1. [CrossRef]
48. Ganeev, R.A.; Boltaev, G.S.; Kim, V.V.; Zhang, K.; Zvyagin, A.I.; Smirnov, M.S.; Ovchinnikov, O.V.; Redkin, P.V.; Wöstmann, M.; Zacharias, H.; et al. Effective high-order harmonic generation from metal sulfide quantum dots. *Opt. Express* **2018**, *26*, 35013–35023. [CrossRef]
49. Ganeev, R.A.; Boltaev, G.S.; Kim, V.V.; Venkatesh, M.; Zvyagin, A.I.; Smirnov, M.S.; Ovchinnikov, O.V.; Wöstmann, M.; Zacharias, H.; Guo, C. High-order harmonic generation using quasi-phase matching and two-color pump of the plasmas containing molecular and alloyed metal sulfide quantum dots. *J. Appl. Phys.* **2019**, *126*, 193103. [CrossRef]

applied sciences

MDPI

Article

Probing Laser Plasma Dynamics Using High-Order Harmonics Generation in Carbon-Containing Nanomaterials

Rashid A. Ganeev [1,2,3,4,*], Vyacheslav V. Kim [1,2], Konda Srinivasa Rao [1] and Chunlei Guo [5]

1 GPL, State Key Laboratory of Applied Optics, Changchun Institute of Optics, Fine Mechanics and Physics, Chinese Academy of Sciences, Changchun 130033, China; vkim@aus.edu (V.V.K.); ksrao@ciomp.ac.cn (K.S.R.)
2 Institute of Astronomy, University of Latvia, LV-1586 Riga, Latvia
3 Department of Physics, Voronezh State University, 394006 Voronezh, Russia
4 Laboratory for the Photonics of Quantum Nanostructures, Moscow Institute of Physics and Technology, 141701 Dolgoprudny, Russia
5 The Institute of Optics, University of Rochester, Rochester, NY 14627, USA; chunlei.guo@rochester.edu
* Correspondence: rashid_ganeev@mail.ru

Abstract: We study high-order harmonics generation from plasmas generated from graphite, fullerenes, carbon nanotubes, carbon nanofibers, diamond nanoparticles, and graphene. Our approach utilizes a heating nanosecond laser pulse to produce plasmas that serve as the media for high harmonic generation from a subsequent driven femtosecond laser pulse. High harmonics are generated at different time delays following the plasma formation, which allows us to analyze the spreading of species with different masses. We analyze the harmonic yields from species of single carbon atom, 60 atoms (fullerene), 10^6 atoms (diamond nanoparticles), 10^9 atoms (CNTs and CNFs), and even much larger species of graphene sheets. The harmonic yields are analyzed in the range of 100 ns–1 ms delays. The harmonic yields were significantly higher within the 200 ns–0.5 μs range, but no harmonic is observed between 10 μs–1 ms. Our observations show that, at the optimal ablation of atoms and clusters, the laser-induced plasmas produced on the surfaces of different carbon-contained species spread out from targets with the comparable velocities.

Keywords: high-order harmonic generation; nanoparticles; carbon plasma

Citation: Ganeev, R.A.; Kim, V.V.; Rao, K.S.; Guo, C. Probing Laser Plasma Dynamics Using High-Order Harmonics Generation in Carbon-Containing Nanomaterials. *Appl. Sci.* **2021**, *11*, 2143. https://doi.org/10.3390/app11052143

Academic Editor: Gennady M. Mikheev

Received: 6 February 2021
Accepted: 24 February 2021
Published: 28 February 2021

1. Introduction

High-order harmonic generation (HHG) in extreme ultraviolet (XUV) range, which occurs when an intense laser pulse interacts with laser-induced plasma (LIP) formed by another laser pulse on the surfaces of bulk targets, is a rich field for a variety of material studies [1–15]. The high-order nonlinear optical properties of various inorganic nanoparticles (Ag, Au, Pd, Pt, Ru, $BaTiO_3$, and $SrTiO_3$) appearing in LIP were reviewed in [16].

The heating pulses (HP) are usually generated by splitting the amplified laser pulse beam (with pulse duration of a few hundred ps) before temporal compression in a chirped-pulse amplification laser system. The time delay between the picosecond HP and the femtosecond driving pulses (DP) for HHG is commonly created by the path length difference between the two optical paths, so the delays are usually restricted by ~100 ns during most of the previous HHG experiments. Correspondingly, the distance between the focus spot and target surface due to above limitations usually does not exceed 200 μm, thus involving particles with minimum velocities distribution at around 10^3 m/s. The presence of clusters consisting of hundreds to thousands of atoms in LIP requires the ability to control delay between HP and DP in a much more comprehensive range, which is difficult while using the optically driven delay between two pulses from the same laser.

One way to overcome this problem is applying a second source, in particular, nanosecond pulses from the Nd: YAG laser. An analysis of the application of longer (nanosecond)

pulses for target ablation during HHG in LIP compared to picosecond pulses, has previously been reported in [17]. Nanosecond laser ablation has been widely used to synthesize various metal and oxide nanoparticles (NPs). However, numerous factors such as the duration of the laser pulse, fluence, and characteristics of irradiated targets affect the properties of synthesized NPs. The pulse duration is one of the most important parameters when considering the in-situ synthesis of NPs in LIP.

In this work, we apply the two-laser scheme to investigate HHG in carbon-based nanostructured plasmas. The nanosecond laser is used for LIP formation and the femtosecond laser is used for harmonics generation in plasma. The synchronization of the two laser sources may resolve, to some extent, the puzzle related to the enhancement of harmonics in the multi-atomic species produced during ablation of bulk materials or targets that initially contain NPs. The analysis of plasma components can be achieved by controlling the HP-DP delay in a wide range spanning between 0 and a few tens of microseconds, which should be sufficient for studying the fast and slow components of LIPs. Such an approach has already been used in reported studies to analyze the harmonics' delay dependencies from different ZnO-contained targets, resonance-enhanced harmonics in mixed LIPs, and HHG during propagation of femtosecond pulses through different ablated species [18–20].

2. Materials and Methods

Femtosecond pulses (50 fs, 806 nm, 2 mJ) from Ti:sapphire laser were focused inside the plasma plume to generate harmonics (Figure 1). The intensity of the driving pulses in the plasma area was maintained to be ~2×10^{14} W cm^{-2}. The beamwaist diameter of the femtosecond driving pulses in the focal plane of the 400 mm focal length spherical lens was measured to be 90 μm and the calculated Rayleigh length was 6 mm, which is notably larger than the sizes of target ablation.

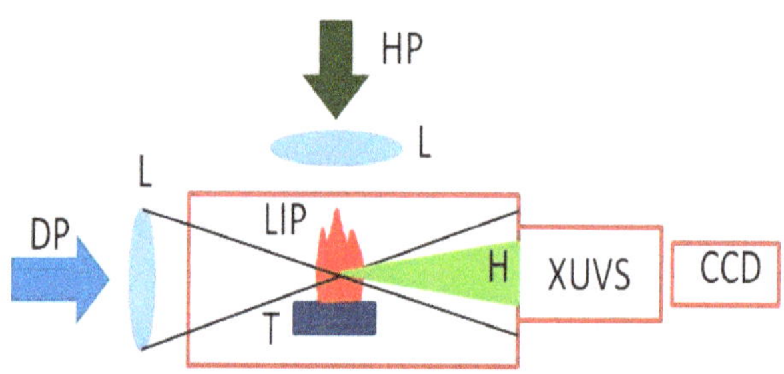

Figure 1. Experimental setup of delayed excitation of nanosecond laser produced plasma for generation of high-order harmonics in carbon nanostructures contained plumes. DP: driving pulse; HP: heating pulse; LIP: laser-induced plasma; T: target; H: harmonic emission; XUVS: extreme ultraviolet spectrometer; CCD: charge-coupled device camera.

For the laser plasma formation, the 5 ns heating laser pulses (λ = 1064 nm, 5 mJ, 10 Hz) from Nd:YAG laser were used. The HP were focused by a 300 mm focal length spherical lens on the target surface at the spot with diameter 0.3 mm providing intensity of a few units of 10^9 W cm^{-2} and fluence up to ~7 J cm^{-2}. The lens used for focusing of heating pulses was approached to the target in such a way that the spot size on the target surface (0.3 mm) was notably larger that the beamwaist diameter in the focal plane (60 μm). The variable defocusing was adjusted to keep the "optimal" fluence of heating radiation for each of used samples, which allowed us to achieve the maximal harmonic yield.

The femtosecond pulses were delayed with regard to those from Nd:YAG laser using the digital delay generator to propagate through the formed plasma at maximal density of the ejected particles. The delay between pulses was tuned using the delay generator. The variable electronic delay range was equal to 0–10^6 ns. We analyzed plasma and harmonic dynamics from several nanoseconds to several hundred microsecond delays.

The harmonics and plasma emission were detected using a flat field grazing-incidence XUV spectrometer.

The pressed carbon-contained nanostructured targets were glued on microscopic glass slides fixed on a digitally controlled XYZ stage inside the target chamber. The powdered nanoparticles were placed in the press machine and pressed to form the 5 mm thick tablets of 10 mm diameter. We used the laser ablation fluence varied between 1 and 7 J cm^{-2} depending on target properties. In each case, the conditions of ablation were chosen to achieve the maximal harmonic yield. The optimization of ablation conditions for each sample assumed a determination of the fluence. The plasma plume allowed the generation of harmonics at a specific delay between heating and driving pulses. The targets were moved around during HHG experiments to provide fresh surface for ablation. Using motorized rotating targets notably improved the stability of HHG from LIPs, particularly, in the case of powdered targets, and significantly minimized the modification of the target surface that could cause degradation of harmonic yield. The periodic change of the ablation zone allowed cooling down the heated area and maintained stable plasma formation. Six materials were used as the targets: bulk graphite (bulk C), C_{60} powder, multi-walled carbon nanotube (MW CNT) powder (15 nm diameter; the length of tubes was varied in the range of 500 nm–20 μm), carbon nanofibers (CNF, 100–200 nm diameter and a few tens micrometers length), diamond nanoparticles (DN, size 3–8 nm) and graphene (GR) wrapped sheets (sizes of sheets ~20 nm). All samples were purchased from Sigma-Aldrich (St. Louis, MO, USA).

3. Results

The first set of our measurements was aimed at the analysis of plasma emission from the ablated species. The plasma emission spectra in the 30–100 nm wavelength range for the six targets are presented in Figure 2.

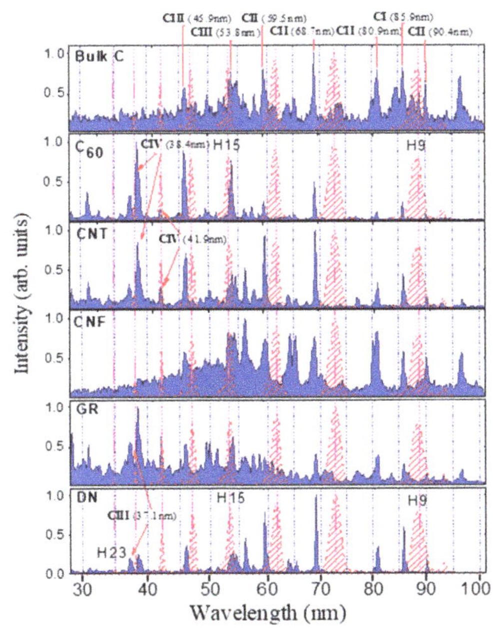

Figure 2. Spectral emission from different carbon-containing plasmas in the case of nanosecond laser ablation. The positions of harmonics (H9–H23) from 806 nm laser source are shown in red.

Plasma emission was recorded at heating pulse parameters (pulse energy and fluence) larger than the optimal ablation characteristics for harmonics generation. The fluence of heating pulses during these studies was equal to 15 J cm^{-2}. We did not average the plasma spectra from these samples but instead used a single shot collection. The emission lines were determined using the NIST database [21]. The positions of the 806 nm radiation harmonics from the bulk carbon starting from 9th (H9) to 23rd (H23) orders are shown as the red curves. The harmonic profiles shown in this figure were included for demonstration of the closeness with some ionic transitions.

For the bulk C plasma produced by HP with the energy of E_{HP} = 3 mJ, we can see the presence of intense CI and CII lines corresponding to the transitions in neutral and singly charged ions of carbon in the spectral region between 55 and 100 nm, as well as the presence of CIII lines below 55 nm. The density of free electrons in bulk C plasma was estimated to be ~5 × 10^{16} cm^{-3}, which is similar to the estimates at the same laser parameters presented in [22]. The particle density was estimated in [23] to be in the range ~10^{18} –10^{19} cm^{-3} for comparable laser fluencies. The contribution of continuum emission decreased the contrast of emission lines.

In the case of the targets comprising nanostructured materials, the shift of emission to larger presence of the lines attributed to CII and CIII transitions was observed, except for CNF and DN. A pattern similar to the above-described spectrum was observed in the case of CNF (E_{HP} = 3 mJ) and GR (E_{HP} = 3 mJ). However, contrary to bulk C plasma, the continuum emission was localized in the shorter wavelength region below 65 nm. In the case of C_{60} (E_{HP} = 2 mJ) and MW CNT (E_{HP} = 2 mJ), the relative intensities of CIII lines significantly increased, while additional strong CIV transitions appeared in the plasma spectra. A similar pattern was observed in the case of all nanostructure-containing plasmas, except CNF. In the case of C_{60}, this behavior confirms the earlier reported observations that, under similar experimental conditions, the plasma from fullerenes reaches much higher ionization states than in the case of bulk graphite ablation [24]. This statement is valid for all nanostructured materials, while at the same time, we observed a decrease in the relative intensity of CI (85.9 nm) line corresponding to the transition in neutral carbon.

In DN (E_{HP} = 4 mJ) case, the plasma spectrum contained less emission lines and a smaller continuum despite the stronger excitation of the target. The different applied HP energies for different samples are explained by the different optimal conditions for plasma formation suitable for HHG. The term "optimal conditions" refers here to the maximal harmonic yield from each ablated sample, which depends on the fluence of HP on the target surface at similar geometry of the focusing conditions. Additionally, the pressed tablets were destroyed at higher fluencies of HP due to their fragility. For example, the harmonic yield from GR and DN LIPs was more than twice as small as from other targets due to weaker ablation at which the pressed tablets remain intact. Correspondingly, we could not increase the HP fluence on those fragile surfaces.

As shown below, the ablation corresponded to the formation of low-ionized plasma using suitable fluence of HPs provided the conditions for efficient HHG. It is important to note that higher densities of free electrons cause the phase mismatch between the driving and harmonics waves for overheated and strongly ionized plasmas, which plays a destructive role in HHG. The moderate fluencies of nanosecond pulses also increased the lifetime of the targets and provided better survival of the original nanostructures being ablated and spread in the plasma plume without their destruction. It was demonstrated that, in the case of laser ablation of C_{60}, MW CNT, CNF, and DN [25] at the experimental conditions similar to ours, the transmission electron microscopy analysis of deposited debris has shown that laser-produced plumes contained the original nanostructures.

Figure 3 shows the harmonic spectra in different plasmas at variable delays between heating and driving pulses in the 100–600 ns range. The dependences of the harmonic yield on the delay are plotted in Figure 4a.

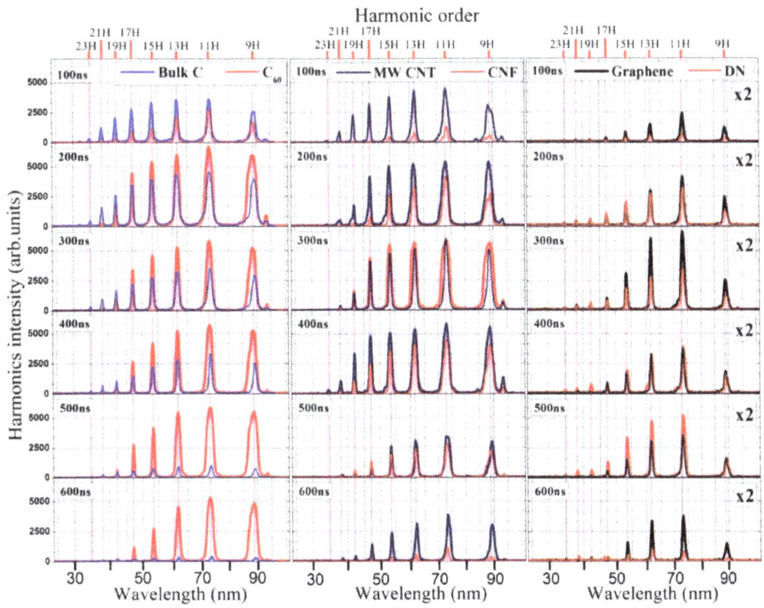

Figure 3. High-order harmonic generation (HHG) spectra generated from the plasmas produced on the bulk C and C_{60} (left column), CNT and CNF (middle column), GR and DN (right column) at the HP-DP delays ranging from 100 ns to 600 ns. The contribution to HHG spectra from different components of ablated plume spreading with almost similar velocities is shown.

The left column of Figure 3 compares the harmonics generated from 9th to 25th order from bulk C (blue curve) and C_{60} (red curve). The maximal harmonic yields for both samples were observed at ~200 ns delay from the beginning of the ablation. Meanwhile, the delay dependence for C_{60} (Figure 4a, red line with empty squares) demonstrates higher integrated harmonics signal, as well as it is extended up to longer delays (~1 μs). In the case of bulk C (Figure 4a, black line with filled squares) we observed a pronounced single maximum at 200 ns, which can be interpreted as the moment of arriving of the largest amount of elemental emitters to the interaction area with DP. The estimation for the velocity of the main part of emitters shows that it equals 1.2×10^3 m/s, which is close to our HHG-based measurements in the case of DP propagation at 0.2 mm distance from the target surface (~(1–2) $\times 10^3$ m/s). The delay dependence in the case of C_{60} (Figure 4a) demonstrates a weak decay between 200 and 600 ns, which can be interpreted as a gradual passing of C_{60} and its different fragments through the interaction area at the velocities slightly varying from 1×10^3 m/s. As one can see (left column of Figure 3), in the case of harmonic spectrum from C_{60} their yield gradually increased and maintained almost equal for the 9th to 17th orders in the range of 200 to 500 ns delays.

Figure 4b shows the saturated images of harmonics collected by CCD camera from the phosphorous screen of the microchannel plate of XUV spectrometer at the optimal delays of each studied sample. The images are shown in log scale and normalized to 1. The purpose of such presentation of harmonics distribution is to define the maximal order of generated harmonics, i.e., harmonic cutoff, and better view the difference in harmonic intensities from various species. In this figure, an emission is observable on the right side of 9H in the first five spectra, which is a second-order diffraction related with strong emission from 17th order. One can also see the weak second-order diffraction related to the 19H (see third panel from the top in the case of MW CNT), which is seen on the left side of 9H. Similar pattern is seen in some panels of Figure 3 showing the lineouts of the harmonic spectra.

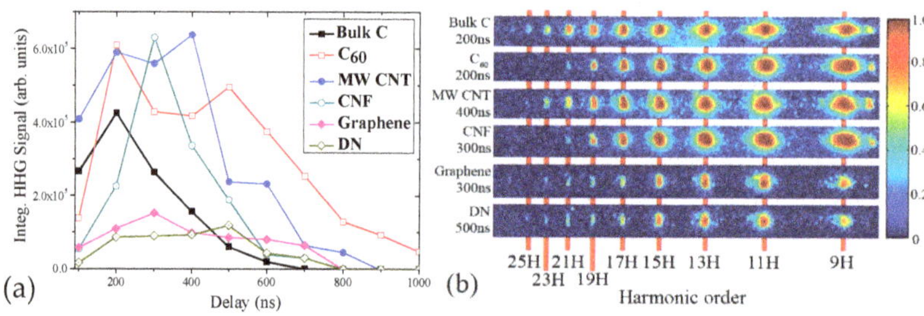

Figure 4. Panel (**a**): dependences of the integrated yield comprising 9th to 25th harmonics on the delay between nanosecond heating pulse (HP) and femtosecond driving pulse (DP). Panel (**b**): raw images of the harmonics at the optimal delay for each plasma sample. Images are normalized and logarithmically scaled. For bulk C, the maximal yield of harmonics is shown at a a delay equal to 200 ns. The optimal delays for C_{60}, MW CNT, CNF, Gr, and DN were 200, 400, 300, 300, and 500 ns, respectively.

For bulk C and C_{60} we notice a difference in the harmonic cutoffs. From all six samples, ablated bulk C allowed the generation of the highest cutoff at similar conditions of experiments (25th order; we also observed very weak 27th harmonic, however, its signal was close to the sensitivity limit of our detection system). Harmonics generating in C_{60} LIP showed the smallest cutoff position (21th order).

4. Discussion

Table 1 comprises the optimal conditions of ablation for different targets. Note that earlier studies have demonstrated that neither pulse energy nor pulse intensity can be considered as the definitive parameters responsible for the optimization of plasma formation to produce maximal yield of harmonics. The most important parameter here is the fluence, or energy density, of the heating pulses on the target surface. In this table, we include the optimal fluencies for each studied sample.

Table 1. Summary of the optimal fluencies of heating pulses for each of studies carbon-containing target.

		F, J cm^{-2}
1	Graphite	3
2	Fullerenes	1.4
3	Carbon nanotubes	1.2
4	Carbon nanofiber	1.2
5	Diamond nanoparticles	1.7
6	Graphene	1.4

It was demonstrated in [23,25,26] by time-of-flight mass spectroscopy (TOFMS) that LIPs from the graphite surface contain, along with neutral carbon atoms and its ions, different C_n clusters and their ions. The number of carbon atoms in such clusters varies from $n = 2$ up to $n = 20$, and for strong excitation n could even reach 100.

Earlier, in the case of C_{60} ablation at conditions similar to ours, TOFMS demonstrated the presence of preferably C_{26}^+ fragments and smaller amount of C_{60}^+ ions, alongside the neutral nanoparticles, in LIP [25]. In order to provide a qualitative explanation of the obtained HHG results we use the semiclassical model of HHG in isotropic medium [27]. According to this model, the HHG cutoff is defined by the expression $N_c \approx (I_p + 3.2U_p)/\omega$, where I_p is the ionization potential of elemental emitter, ω is the laser field frequency, $U_p = I_l/(4\omega^2)$ is the pondermotive potential, and I_l is the laser field intensity. The ionization potential of carbon atom is I_p (Bulk C) = 11.26 eV [28], while the ionization potential for small-sized carbon molecules varies from I_p (C_2) = 11.35 eV, I_p (C_3) = 11.5 eV to I_p (C_9) = 9.4 eV [26] with the tendency to decrease toward bigger molecules. For C_{60}, the ionization potential is correspondingly lower compared to smaller-sized molecules (I_p (C_{60}) = 7.6 eV [29]). The application of our experimental parameters in expression for harmonic cutoff gives N_c (bulk C) = 31 and N_c (C_{60}) = 29. Note that in the present work the 25th and 21st harmonics cutoffs from those species were observed. Thus, we attribute the observed peculiarities of HHG spectra in pair "bulk C–C_{60}" to the presence of mostly C_{60} and its large fragments in the plasma plume, which was indirectly confirmed by our plasma emission spectra (Figure 2, C_{60} panel), where the relative intensity of the CI transition line is lower than that of CIII and CIV lines. Larger harmonics cutoff in bulk C targets can be attributed to the presence of mostly neutral and small-sized carbon molecules in LIP. The presence of highly charged ions does not provide sufficient contribution to the extension of HHG cutoff, since the HHG conversion efficiency quickly reduces with increasing ionization potential.

Returning to the model [27], harmonics generation is described as a process where the first step is the formation of the electron wave packet by tunneling ionization and the second step includes a classical mechanics description of released electron in laser field. The first step can be described by the Ammosov–Delone–Krainov (ADK) expression for ionization rates Γ_{ion} for different ionization potentials [30] and defines in two-step model the efficiency of conversion. While estimating ADK rates we assumed the laser intensity to be equal to what we measured in the focal plane of the focused driving beam (2×10^{14} W cm^{-2}). Direct comparison of ionization rates by ADK formula gives $\Gamma_{ion}[I_p(C) = 11.26 \text{ eV}]/\Gamma_{ion}[I_p(C^+) = 24.38 \text{ eV}]$~$10^5$ higher rate for tunnel ionization in the cases of the neutral and single charged carbon atom. By applying the same approach to evaluate the difference in ionization rates of fullerenes and carbon, one can define the following ratio $\Gamma_{ion}[I_p(C_{60}) = 7.6 \text{ eV}]/\Gamma_{ion}[I_p(C) = 11.26 \text{ eV}] \approx 5$.

Thus, C_{60} molecules can be considered more efficient high harmonics emitters than ordinary carbon atoms or small-sized carbon molecules. Their presence in LIP leads to a stronger enhancement of the low-order part of HHG spectrum comprising the harmonics between 9th and 17th orders. We can conclude that, despite the lower density of C_{60} nanoparticles in plasma compared to C atoms and ions originated from graphite ablation, higher HHG efficiency allows the observation of harmonics even at notably larger delays (>1000 ns) compared with the latter species (Figure 4a). Higher cutoff position in the case of ablated graphite is attributed to the presence of neutral carbon and small-sized C_n (where $n < 20$) molecules. Note that the role of different small carbon clusters synthesized during ablation of graphite was analyzed in the studies related to HHG in carbon LIP [20].

The middle column of Figure 3 represents HHG spectra obtained from, to some extent, morphologically similar species: multiwalled carbon nanotubes (blue solid lines) and carbon nanofibers (red solid lines). MW CNT plasma dynamics showed the 400 ns optimal delay from the beginning of ablation corresponding to maximal harmonic yield. Simultaneously, the delay dependence (Figure 4a, blue line with solid circles) had the plateau-like structure between 200 and 400 ns. The harmonic yield in this time scale was equal or slightly higher than in the case of C_{60} plasma. The enhanced harmonics from 9th to 17th orders were observed between 100 and 400 ns delays (Figure 3, middle column). At the same time, the cutoff (25th harmonic) at the optimal delay (400 ns) was slightly higher compared to C_{60} plasma (Figure 4b).

We can assume that the enhanced low-order harmonics were attributed to the presence of small fragments of CNTs in LIP. At the same time, higher cutoff can be attributed to the coherent contribution to HHG from the carbon neutral atoms and small-sized carbon molecules. Notice that the ionization potential of single-walled CNT is in the range of 5–6 eV [31]. The HHG spectrum in the case of short (100 ns) delay indicates that small-sized components of such plasma can gain, to some extent, larger velocities, thus causing the enhanced harmonic spectra with larger cutoff order. This is what we see in the case of bulk C and MW CNT. At the same time, the large-sized components of LIP responsible for enhancing the lower orders of spectra are expected to appear at larger delays. Summarizing the analysis of HHG in MW CNT, one can conclude that, at the used parameters of HP, MW CNT plasma also contains small carbon particles alongside the fragments of nanotubes.

Meanwhile, the HHG spectrum from CNF was surprisingly similar to C_{60}. CNF plasma showed optimum delay at 300 ns (Figure 4a, green line with empty circles) and similar to C_{60} cutoff (21st order, Figure 4b). We assume that plasma from CNF ablation was less fragmented compared to MW CNT despite the higher HP energy used there (3 mJ vs. 2 mJ).

Earlier reported TOFMS spectra have demonstrated the similarities between the MW CNT and CNF, with the presence of small carbon C_n fragments (with n varied up to 20) and no ions higher than low-sized carbon clusters in the mass spectra up to 5000 mass/charge units [25]. Additionally, the emission from ablated MW CNT and C_{60} targets showed almost similar spectra (Figure 2), with strong relative lines for CIII and CIV transitions. We can suppose that, in the case of MW CNT, the fragmentation process induces the appearance of small-sized carbon molecules and clusters inducing the larger cutoff, since at short delay (100 ns, Figure 4a) the integrated harmonic signal from MW CNT plasma exceeds that from bulk C LIP. Plasma spectra for CNF demonstrate the most contradictive result, with strong relative CI transition line, which indirectly points out to the presence of large number of neutral carbon atoms, but without extending the HHG cutoff.

TOFMS is a natural and obvious way to precisely determine the mass-to-charge state of the ablation plume. Notice that the main goal of present research was to study high-order harmonics generation from the plasmas generated on the graphite, fullerenes, carbon nanotubes, carbon nanofibers, diamond nanoparticles, and graphene targets. Meantime, the joint application of HHG and TOFMS facilities in a single set of experiments is hardly to be achieved. Earlier studies, which combined the data revealed from above facilities, were performed separately, which did not allow for a conclusive argument for the similarity of the plasma formation conditions in these two sets of experiments. Meanwhile, we demonstrate that the HHG approach allows for analyzing the dynamics of plasma spreading out from ablated surface thus revealing some interesting peculiarities hardly obtainable in the case of TOFMS approach. Our present research, as well as earlier reported studies of harmonic generation in laser-induced plasmas, enabled a demonstration of the attractive features of the high-order nonlinear spectroscopy of these species.

The HHG spectra generated in the plasmas produced on the graphene sheets and diamond nanoparticles are shown on the right column of Figure 3. The harmonic yields from GR and DN was multiplied by a factor of 2 to make them more visible and, in order to compare with other spectra. Both plasmas allowed the generation of weaker harmonics compared to the other four samples. The delay dynamics in the case of GR (Figure 4a, solid pink line with filled rhombuses) indicated the presence of large fragments in LIP, with maximal harmonic yield observed at 300 ns delay. The delay dynamics curve is more flattened with respect to the above-discussed materials, with comparable integrated harmonic yield in the range of 100 ns to 700 ns. We attribute this pattern of delay dynamics to the wide distribution of GR's pieces possessing different velocities centered at ~1×10^3 m/s. The harmonic spectrum changed at optimum delay (Figure 4b) and showed the cutoff at 23rd order, and the enhanced 11th and 13th harmonics (Figure 3, left column, black line). We attribute this increase in harmonic yield to the arrival of large-sized fragments of graphene to the interaction area.

Among all studied materials, DNs have the largest crystal lattice energy. This characteristic can explain, to some extent, the better survival of DNs in the plasmas produced by the heating pulses. This may also explain why DNs showed relatively extended optimum delay (500 ns; Figure 4a, dark yellow line with empty rhombuses). With the increasing of the delay, the 11th to 15th harmonics also enhanced (Figure 3, right column, red solid lines), while up to 500 ns delay the harmonic spectral shape and cutoff located at 25th order remained the same (Figure 4b). The observed peculiarities of HHG spectra from GR and DN plasmas can be explained by the presence of the small-sized carbon molecules and clusters having uniform distribution of velocities. The latter assumption can explain the flattened delay curves with the same HHG cutoffs (23rd and 25th orders).

The cutoff positions for all studied samples were, to some extent, close to each other except for graphene (Figure 3). Though the ionization potentials were almost similar, the studied plasmas were distinguished from each other by the harmonic yield rather than by cutoff energy.

Below, we address the importance of considering the above delay-dependent experiments for understanding the dynamics of plasma propagation. The signature, indicating the presence of specific emitters, is the increase in the output of harmonics at certain specific delays from the beginning of ablation. We can assume that the same average kinetic energy $E = mv^2/2$ could characterize all plasma components containing the same basic atom, in our case carbon. Accordingly, the same average arrival time could be expected for carbon clusters of different sizes. The duration for cluster propagation from the target surface to the optical axis of the femtosecond pulse propagation corresponds to the optimal delay between HP and DP. Our results show that small carbon clusters enter the area of interaction with the femtosecond beam much earlier than expected, assuming only kinetic consideration when it is assumed that the components reach thermodynamic equilibrium during expansion. Alternatively, we can suppose that all carbon clusters acquire the same kinetic energy from the very beginning and propagate from the surface at speed approximately similar to that of a single carbon ablated from bulk material. The arrival times of particles consisting of a single carbon atom, 60 atoms (fullerene), 10^6 atoms (diamond nanoparticles), 10^9 atoms (CNT and CNF), and even much larger species, such as parts of graphene sheets, were comparable to each other (200-400 ns). Thus, the role of the cluster's atomic weight in HHG becomes questionable if only a simple kinetic energy mechanism is considered. In other words, from the very beginning, all the atoms in the clusters acquire the same kinetic energy and propagate from the surface at speed approximately equal to that of a single carbon atom.

When we use the term "heating" for the laser ablation, we do not merely suppose the thermal evaporation of the target. It is not a simple thermal evaporation induced plasma formation. This process depends on the level of target excitation that is used for laser ablation. The creation of nonlinear medium above the target surface is not based on the simple heating of the target surface and steady-state processes of melting, evaporation, and spreading of the particles with the velocities defined from the thermodynamic relations. This relation refers to cw heating. In this case the velocity of the C_{60} molecule at 1000 K is in the range of 1.5×10^2 m s^{-1}. Correspondingly, during the first few hundred nanoseconds (~200 ns corresponding to our observation of maximal yield of harmonics) the fullerene molecules will move only 30 μm above the surface. Notice that femtosecond driving pulse propagates at the distance of ~0.2 mm above the target's surface. If one assumes that plasma creation by laser pulses is defined by this slow process, then no harmonics at all should be observed in such experimental configuration for any of our targets, be it C_{60}, CNT, CNF, graphene, or DN. In the meantime, laser ablation of any of the above targets creates a very efficient plasma medium, which generates extreme high harmonics when the femtosecond pulse propagates 200 μm above the target surface.

This contradiction is explained by another model of creation of the cloud of particles, namely, plasma explosion during ablation of the targets. The dynamics of plasma front propagation during laser ablation is studied by few groups (for example [32] and references therein). The dynamics of plasma formation and spreading can be analyzed by either the time resolved ICCD images or shadowgram technique. A numerical analysis of the generation of such plasmas for the case of single-pulse interaction with the target surface was described in [33]. Previously, the dynamics of the spatial characteristics of laser plasmas generated from B and Mo targets, measured using the shadowgraphs of the plasma, was reported in [34]. For Mo, the plasma front spreads with the velocity of ~6×10^3 m s^{-1}. For example, the plasma front reaches 200 μm distance from the target after hundred nanoseconds rather than a few thousand ns, as we can estimate if we assume the steady-state expansion of the plasma particle cloud. Obviously, the formation of "optimal" plasma is not restricted by appearance of the plasma front in the area of femtosecond pulse. We must wait until the density of the particles becomes sufficient for efficient HHG, while the free-electron concentration remains below the level when the impeding processes prevail over the harmonic generation. In our case the velocity of the "optimal" part of carbon nanostructures contained plasma cloud was measured to be 1×10^3 m s^{-1}.

The appearance of continuum points out the strong heating of target surface. The strong incoherent emission of plasma also decreases the overall "quality" of generated XUV radiation when the former emission becomes stronger than coherent emission of harmonics. That is why the determination of "optimal" plasma for HHG [35] includes various factors playing an important role in amendment and suppression of harmonic yield. Briefly, moderate (10^{17} cm^{-3}) plasma density, a small ionization rate, and correspondingly small electron density (a few units of 10^{16} cm^{-3}) helped achieve the best conversion efficiencies towards the harmonics in the plateau range (10^{-5} and higher), which was almost two orders of magnitude higher than in the case of "overheated" targets. The application of "optimal" plasma immediately allowed for the demonstration of such unique properties of HHG in ablated debris like the resonance-induced enhancement of single and group of harmonics, the resonance-induced suppression of some harmonics, the quasi-phase matching of the groups of harmonics in different ranges of XUV, the efficient application of clusters, quantum dots and relatively large nanoparticles for harmonics generation, the application of extended plasmas at the conditions when the coherence length for some harmonics does not exceed the sizes of laser-induced torches, etc. These advanced features of HHG in optimally prepared plasmas cannot be reproduced in the plasmas formed during strong ablation of target surfaces. Moreover, most of above advantages could not be repeated in the gas media commonly used of HHG.

The continuum shown in the emission spectra was produced during the strong excitation of our targets. The motivation to show plasma emission spectra at these conditions is: (a) to demonstrate the similarity in emission lines of different nanostructured materials caused by the presence of carbon in each of these species and (b) to show the closeness of harmonic and plasma emission lines, which does not lead to the resonance-induced enhancement of those harmonics. Moreover, our spectral measurements were carried out at the conditions when we could not distinguish the emission from the target surface and emission from nearby (~0.2 mm above the ablated surface) plasma.

At the moderate ablation conditions (i.e., at the fluence of 2 to 7 J cm^{-2} on the target surface) the nanostructures are not affected by the temperature of surrounding material (~900 K) but rather safely elevate from the surface. The survival of carbon nanostructures at these conditions of ablation and their presence at the moment of propagating the driving femtosecond pulses through such plasma formations was confirmed by the SEM analysis of the debris deposited on the nearby surfaces.

To summarize, we considered two models: the model of plasma spreading and the model of harmonics generation in the plasmas comprising of the same element (carbon, in our case) either in a single atomic mode or complex morphology state. The first model allowed us to confirm the similarity in the velocities of single-atomic and multi-atomic debris, which corroborates with our observations of the maximal yield of harmonics at approximately same delay between the heating and driving pulses for each of these species. The second model offers the consideration of C and C_n (n refers to the number of atoms in cluster) particles as the harmonic emitters from the point of view of their ionization potentials and harmonic yield. An approximate similarity in the ionization potentials led to similarity, to some extent, of the cutoff harmonics, while a larger number of atoms in nanostructures allowed for increasing the cross section of recombination of the accelerated electron with the parent particle. The latter peculiarity distinguished the harmonic yields from the single-atomic plasma produced from ablated graphite and the multi-atomic plasma produced from other studied carbon-contained species (fullerenes, multiwalled nanotubes, nanofibers, diamond nanoparticles, and graphene). Thus, by combining the analysis of delay dependences of the integrated harmonic signals and HHG spectra in the 30–100 nm wavelength range, the roles played by the carbon monomers, small-sized carbon molecules, and large nanoparticles in the former process were revealed. A search of new approaches in this direction allows for demonstrating the advantages of high-order nonlinear spectroscopy of ablated nanostructured solids.

5. Conclusions

In conclusion, we presented studies on the high-order harmonics generation in the plasmas generated from six carbon-contained materials (graphite, fullerenes, multiwalled nanotubes, nanofibers, diamond nanoparticles, and graphene). The contribution of different components in laser ablated plume from carbon nanostructure-containing targets in the range of 100 ns to 1 μs delays between HP and DP was analyzed. We have shown the difference of delay dependence curves between bulk graphite and other samples containing nanoparticles while recognizing the similarity of the optimal values of delays for different species at which highest harmonic yield was achieved. By combining the analysis of delay dependences of integrated harmonics signals and of HHG spectra in the 30–100 nm wavelength range, the role of carbon monomers, small-sized carbon molecules, and large nanoparticles in the former process were revealed. We demonstrated that the presence of nanoparticles and its fragments leads to an enhanced harmonics signal in the wavelength range of 50–95 nm (i.e., harmonics from 9th to 17th orders), especially in C_{60}, CNF and MW CNT. It was also shown that different carbon-containing species demonstrate sufficiently variable patterns of HHG spectra at comparable heating pulse energy. The important role of utilizing two digitally controlled laser sources for obtaining extensive delays between heating and driving pulses was underlined.

The presented studies allow for a better understanding of the plasma spreading dynamics in the harmonics generation from ablated species comprising similar basic element (carbon). The approximately similar average arrival times were observed for carbon clusters of different sizes. The arrival times for the particles comprising single carbon atom, 60 atoms (fullerene), 10^6 atoms (diamond nanoparticles), 10^9 atoms (CNTs and CNFs), and even much larger species like parts of graphene sheets, were comparable (200–400 ns). We analyzed the HHG yield in the range of 10 μs–1 ms delays and did not observe any harmonics. Our observations showed that, at the optimal ablation of atoms and clusters, the laser-induced plasmas produced on the surfaces of different carbon-contained species spread out from targets with the comparable velocities.

Author Contributions: Conceptualization, R.A.G., C.G.; methodology, R.A.G., V.V.K.; formal analysis, V.V.K., K.S.R.; investigation, V.V.K., K.S.R.; writing—original draft preparation, V.V.K., R.A.G.; writing—review and editing, R.A.G., C.G.; visualization, V.V.K.; supervision, C.G.; funding acquisition, R.A.G., C.G. All authors have read and agreed to the published version of the manuscript.

Funding: This research was funded by Jilin Provincial Science & Technology Development Project (20200802001GH), K. C. Wong Education Foundation (JTD-2018–08), ERDF project (1.1.1.5/19/A/003), State Assignment to Higher Educational Institutions of Russian Federation (FZGU-2020-0035).

Institutional Review Board Statement: Not applicable.

Informed Consent Statement: Not applicable.

Data Availability Statement: The data that support the findings of this study are available from the corresponding author upon reasonable request.

Conflicts of Interest: The authors declare no conflict of interest.

References

1. Ganeev, R.A.; Suzuki, M.; Kuroda, M.B.A.H. Harmonic generation in XUV from chromium plasma. *Appl. Phys. Lett.* **2005**, *86*, 131116. [CrossRef]
2. Ganeev, R.A.; Singhal, H.; Naik, P.A.; Arora, V.; Chakravarty, U.; Chakera, J.A.; Khan, R.A.; Redkin, P.V.; Raghuramaiah, M.; Gupta, P.D. Single harmonic enhancement by controlling the chirp of the driving laser pulse during high-order harmonic generation from GaAs plasma. *J. Opt. Soc. Am. B* **2006**, *23*, 2535–2540. [CrossRef]
3. Ozaki, T.; Bom, L.B.E.; Ganeev, R.; Kieffer, J.-C.; Suzuki, M.; Kuroda, H. Intense harmonic generation from silver ablation. *Laser Part. Beams* **2007**, *25*, 321–327. [CrossRef]
4. Ganeev, R.A.; Singhal, H.; Naik, P.A.; Chakravarty, U.; Arora, V.; Chakera, J.A.; Khan, R.A.; Raghuramaiah, M.; Kumbhare, S.R.; Kushwaha, R.P.; et al. Optimization of the high-order harmonics generated from silver plasma. *Appl. Phys. B* **2007**, *87*, 243–247. [CrossRef]
5. Ganeev, R.A. Generation of high-order harmonics of high-power lasers in plasmas produced under irradiation of solid target surfaces by a prepulse. *Phys. Uspekhi* **2009**, *52*, 55–77. [CrossRef]
6. Ganeev, R.A.; Bom, L.B.E.; Wong, M.C.H.; Brichta, J.-P.; Bhardwaj, V.R.; Redkin, P.V.; Ozaki, T. High-order harmonic generation from C_{60}-rich plasma. *Phys. Rev. A* **2009**, *80*, 043808. [CrossRef]
7. Hutchison, C.; Ganeev, R.A.; Witting, T.; Frank, F.; Okell, W.A.; Tisch, J.W.G.; Marangos, J.P. Stable generation of high-order harmonics of femtosecond laser radiation from laser produced plasma plumes at 1 kHz pulse repetition rate. *Opt. Lett.* **2012**, *37*, 2064–2066. [CrossRef] [PubMed]
8. Ganeev, R.A.; Hutchison, C.; Witting, T.; Frank, F.; Okell, W.A.; Zaïr, A.; Weber, S.; Redkin, P.V.; Lei, D.Y.; Roschuk, T.; et al. High-order harmonic generation in graphite plasma plumes using ultrashort laser pulses: A systematic analysis of harmonic radiation and plasma conditions. *J. Phys. B* **2012**, *45*, 165402. [CrossRef]
9. Singhal, H.; Naik, P.A.; Kumar, M.; Chakera, J.A.; Gupta, P.D. Enhanced coherent extreme ultraviolet emission through high order harmonic generation from plasma plumes containing nanoparticles. *J. Appl. Phys.* **2014**, *115*, 033104. [CrossRef]
10. Ganeev, R.A.; Suzuki, M.; Kuroda, H. Quasi-phase-matching of high-order harmonics in multiple plasma jets. *Phys. Rev. A* **2014**, *89*, 033821. [CrossRef]
11. Wöstmann, M.; Redkin, P.V.; Zheng, J.; Witte, H.; Ganeev, R.A.; Zacharias, H. High-order harmonic generation in plasmas from nanoparticle and mixed metal targets at 1-kHz repetition rate. *Appl. Phys. B* **2015**, *120*, 17–24. [CrossRef]
12. Fareed, M.A.; Strelkov, V.V.; Thiré, N.; Mondal, S.; Schmidt, B.E.; Légaré, F.; Ozaki, T. High-order harmonic generation from the dressed autoionizing states. *Nat. Commun.* **2017**, *8*, 16061. [CrossRef]
13. Fareed, M.A.; Strelkov, V.V.; Singh, M.; Thiré, N.; Mondal, S.; Schmidt, B.E.; Légaré, F.; Ozaki, T. Harmonic generation from neutral manganese atoms in the vicinity of the giant autoionization resonance. *Phys. Rev. Lett.* **2018**, *121*, 023201. [CrossRef] [PubMed]
14. Wöstmann, M.; Splitthoff, L.; Zacharias, H. Control of quasi-phase-matching of high-harmonics in a spatially structured plasma. *Opt. Express* **2018**, *26*, 14524–14537. [CrossRef]
15. Kumar, M.; Singhal, H.; Chakera, J.A. High order harmonic radiation source for multicolor extreme ultraviolet radiography of carbon plumes. *J. Appl. Phys.* **2019**, *125*, 155902. [CrossRef]
16. Ganeev, R.A. *Nanostructured Nonlinear Optical Materials: Formation and Characterization*; Elsevier: Amsterdam, The Netherlands, 2018.
17. Ganeev, R.A.; Suzuki, M.; Baba, M.; Kuroda, H. High harmonic generation from the laser plasma produced by the pulses of different duration. *Phys. Rev. A* **2007**, *76*, 023805. [CrossRef]
18. Venkatesh, M.; Ganeev, R.A.; Rao, K.S.; Boltaev, G.S.; Zhang, K.; Srivastava, A.; Bindra, J.K.; Singh, S.; Kim, V.V.; Maurya, S.K.; et al. Influence of gadolinium doping on low- and high-order nonlinear optical properties and transient absorption dynamics of ZnO nanomaterials. *Opt. Mater.* **2019**, *95*, 109241. [CrossRef]
19. Boltaev, G.S.; Ganeev, R.A.; Strelkov, V.V.; Kim, V.V.; Zhang, K.; Venkatesh, M.; Guo, C. Resonance-enhanced harmonics in mixed laser-produced plasmas. *Plasma Res. Express* **2019**, *1*, 035002. [CrossRef]
20. Ganeev, R.A.; Boltaev, G.S.; Zhang, K.; Maurya, S.K.; Venkatesh, M.; Yu, Z.; Kim, V.V.; Redkin, P.V.; Guo, C. Role of carbon clusters in high-order harmonic generation in graphite plasmas. *OSA Continuum* **2019**, *2*, 1510–1523. [CrossRef]
21. Kramida, A.; Ralchenko, Y.; Reader, J.; NIST ASD Team. *NIST Atomic Spectra Database*; (version 5.8); National Institute of Standards and Technology: Gaithersburg, MD, USA, 2020. Available online: https://physics.nist.gov/asd (accessed on 27 February 2021). [CrossRef]
22. Balki, O.; Rahman, M.M.; Elsayed-Ali, H.E. Optical emission spectroscopy of carbon laser plasma ion source. *Opt. Commun.* **2018**, *412*, 134–140. [CrossRef]
23. Creasy, W.R.; Brenna, J.T. Large carbon cluster ion formation by laser ablation of polyimide and graphite. *Chem. Phys.* **1988**, *126*, 453–468. [CrossRef]
24. Wülker, C.; Theobald, W.; Ouw, D.; Schäfer, F.P.; Chichkov, B.N. Short-pulse laser-produced plasma from C60 moleules. *Opt. Commun.* **1994**, *112*, 21–28. [CrossRef]
25. Ganeev, R.A.; Baba, M.; Suzuki, M.; Kuroda, H. Morphology of laser-produced carbon nanoparticle plasmas and high-order harmonic generation of ultrashort pulses in clustered media. *J. Phys. B At. Mol. Opt. Phys.* **2014**, *47*, 135401. [CrossRef]
26. Weltner, W.; van Zee, R.J. Carbon molecules, ions, and clusters. *Chem. Rev.* **1989**, *89*, 1713–1727. [CrossRef]

27. Corkum, P.B. Plasma perspective on strong field multiphoton ionization. *Phys. Rev. Lett.* **1993**, *71*, 1994–1997. [CrossRef]
28. Haris, K.; Kramida, A. Critically evaluated spectral data for neutral carbon. *Astrophys. J. Suppl. Ser.* **2017**, *233*, 16. [CrossRef]
29. Ding, D.; Compton, R.N.; Haufler, R.E.; Klots, C.E. Multiphoton ionization of C60. *J. Phys. Chem.* **1993**, *97*, 2500–2504. [CrossRef]
30. Ammosov, M.V.; Delone, N.B.; Krainov, V.P. Tunnel ionization of complex atoms and of atomic ions in an alternating electromagnetic field. *Sov. Phys. JETP* **1986**, *64*, 1191–1194.
31. Buonocore, F.; Trani, F.; Ninno, D.; di Matteo, A.; Cantele, G.; Iadonisi, G. *Ab initio* calculations of electron affinity and ionization potential of carbon nanotubes. *Nanotechnology* **2008**, *19*, 025711. [CrossRef] [PubMed]
32. Rus, B.; Zeitoun, P.; Mocek, T.; Sebban, S.; Kálal, M.; Demir, A.; Jamelot, G.; Klisnick, A.; Králiková, B.; Skála, J.; et al. Investigation of Zn and Cu prepulse plasmas relevant to collisional excitation x-ray lasers. *Phys. Rev. A* **1997**, *56*, 4229–4236. [CrossRef]
33. Hora, H. *Plasmas at High Temperature and Density*; Springer: Berlin/Heidelberg, Germany, 1991.
34. Ganeev, R.A.; Suzuki, M.; Baba, M.; Kuroda, H. Investigation of boron and molybdenum plasma at the multi-pulse interaction of femtosecond radiation with target. *Opt. Spectrosc.* **2005**, *99*, 1000–1005. [CrossRef]
35. Ganeev, R.; Suzuki, M.; Baba, M.; Kuroda, H.; Ozaki, T. High-order harmonic generation from boron plasma in the extreme-ultraviolet range. *Opt. Lett.* **2005**, *30*, 768–770. [CrossRef] [PubMed]

applied sciences

MDPI

Article

Nonlinear Optical Microscopy of Interface Layers of Epitaxial Garnet Films

Anton Maydykovskiy [1,†], Marina Temiryazeva [2,†], Alexey Temiryazev [2,†] and Tatiana Murzina [1,*,†]

1 Department of Physics, Moscow State University, 119991 Moscow, Russia; anton@shg.ru
2 Kotel'nikov Institute of Radioengineering and Electronics of RAS, 141190 Fryazino, Russia; mtemiryazeva@gmail.com (M.T.)
* Correspondence: murzina@mail.ru
† These authors contributed equally to this work.

Abstract: The structure of magnetic domains is an exciting research object that shows an enormous variety of delightful patterns. Epitaxial garnet is one of the most studied magnetic dielectrics with well-recognized bulk domains, while the magnetic composition at the surface is less investigated. Here we apply the nonlinear optical microscopy technique for the visualization of the interface magnetic domains of 10 μm thick $(LuBi)_3Fe_5O_{12}$ film and prove that it is qualitatively similar for both garnet/air and garnet/substrate interfaces. As an efficient extension of the second harmonic generation microscopy, we suggest and demonstrate the possibilities of the third harmonic generation one, which provides higher resolution of the method.

Keywords: magnetic force microscopy; garnet films; magnetic domain structure; optical second harmonic generation

Citation: Maydykovskiy, A.; Temiryazeva, M.; Temiryazev, A.; Murzina, T. Nonlinear Optical Microscopy of Interface Layers of Epitaxial Garnet Films. *Appl. Sci.* **2023**, *13*, 8828. https://doi.org/10.3390/app13158828

Academic Editor: Andrés Márquez

Received: 18 June 2023
Revised: 27 July 2023
Accepted: 27 July 2023
Published: 31 July 2023

1. Introduction

Magnetic domains are of high interest to researchers due to both the bright underlying physics that governs their organization, and because of an important role played by domains in a lot of applications of magnetic nano- and microstructures [1]. One of the most studied and attractive objects here are the crystalline ferrimagnetic iron garnet films that combine strong magnetooptical (MO) activity with high transparency in the red and infrared spectral ranges, pronounced mechanical properties, and the ability to control the parameters of the MO spectra by modification of their chemical composition [2–5]. It is generally recognized that garnets are widely used as optical disks [6], resonators [7], and in microwave devices [8]. Moreover, they are known as an efficient platform for spintronic and magnonic devices due to their exciting ability for the transport and processing of spin waves [9–11].

The most important properties of garnets are determined to a large extent by their domain organization, which is governed by many parameters such as the crystallographic symmetry, type and concentration of dopants, magnetic and crystalline anisotropy, demagnetization fields, etc. [1,2,12,13]. As the properties of the surface and bulk magnetic domains are interconnected, their study is an interesting research task. Furthermore, the development of efficient experimental methods for the visualization of magnetic domains in the surface and bulk areas of garnet films seems also to be quite important.

The best-known approaches here are the polarization-sensitive magnetooptical microscopy typically performed in the Faraday or Voigt geometries [14,15], X-ray diffraction techniques, first of all the surface-sensitive Lorentz microscopy [16,17], and magnetic force microscopy (MFM) [18,19]. In the case of yttrium iron garnet films with perpendicular magnetic anisotropy, the Lorentz microscopy allows to study the magnetic domains at the nanoscale. It also revealed an important role played by the demagnetization field in the formation if domains and domain walls dynamics [20]. These studies involved the high spatial resolution of the Lorentz microscopy technique (that is below 100 nm), and the temporal resolution of this surface sensitive probe can be made less than 1 ps [21]. This allows the study of spin relaxation dynamics and the domain wall's motion under the application of the external magnetic field. The magnetic force microscopy technique allows the visualization of complicated fractal-like and closure domains at the surface of epitaxial garnet films that are absent in bulky materials; it was recognized that the shape of the surface domains depend on the films' thickness and composition [22].

A special place is occupied by the optical second harmonic generation (SHG) [23], which has been shown to provide giant values of magnetooptical effects in the SHG response predicted prior to their experimental observation [24,25]. It is also known for its high surface sensitivity if being applied to materials with intrinsic inversion symmetry [26–30]. In the case of iron garnet films, the application of the SHG technique is beneficial due to (i) symmetry forbiddance of the even-order nonlinear optical effects (including SHG) in the bulk of centrosymmetric materials such as yttrium iron garnet crystals (in the electric dipole approximation), which provides high SHG sensitivity to the properties of the interface regions with inevitably broken inversion symmetry [23,31,32], (ii) absence of the SHG contribution from the substrates, which is typically centrosymmetric gallium-gadolinium garnet (GGG) of a few hundreds of microns in thickness and which can also reveal weak MO responses, (iii) the large value of the MO effects in the SHG response and (iv) larger (as compared to the linear MO microscopy) SHG spatial resolution originating from the nonlinear nature of the effect, and (v) the ability to use a multiple-color pump and probe experimental schemes. As an example, magnetic domains were visualized via the SHG microscopy in [33] in the scheme of the magnetooptical Faraday effect in iron garnet films of different crystallographic orientations, which provided additional information on the domain's magnetization distribution as compared to a linear magnetooptical probe. Symmetry analysis of the magnetic nonlinear optical response at the SHG wavelength was discussed in detail in [34,35]. It was underlined that high-rank tensors describing different components of the SHG process that are of the first and of second orders in magnetization **M**, as well as induced by gradients of **M**, can be used for a thorough characterization of the residual magnetic structure of garnets and other objects.

Third-order nonlinear optical spectroscopy is also a well recognized and powerful technique [36] that is widely applied to the studies of biological objects and tissues [37,38], resonant and local field effects in non-magnetic nanostructured objects [39] and metasurfaces [40,41]. One of substantial benefits of this probe is an intrinsic usage of the two-color scheme. In that case, no laser-induced damage is expected if a tissue is transparent to the intense laser radiation of the infrared spectral range, while a much less intense third harmonic of visible or UV radiation and which is specific to a concrete type of a tissue, enables label-free visualization of biological structures on the scales down to a cell size. At the same time, to the best of our knowledge, magnetization-induced third-order nonlinear optical effects, including third harmonic generation (THG), have not been applied to the studies of magnetic microstructures. This can be due in part to the low values of the magnetization-induced effects in THG as compared to the case of SHG, as was demonstrated in [31] for the case of metallic magnetic nanostructures.

Recently it was shown that the SHG microscopy can distinguish stripe domains in the bulk of garnet films, as well as the closure domains at the outer garnet/air interface [42]. At the same time, the domain structure of the hidden interface of the ferrimagnetic film with the substrate remained unexplored. In this paper we apply magnetization-sensitive nonlinear optical microscopy to the study of magnetic ordering at the garnet/GGG interface and show that it reveals a complicated domain structure similar to that of the garnet/air one. Another finding is the extension of the magnetization-sensitive nonlinear optical microscopy method to the case of third harmonic generation, which shows higher resolution as compared to the case of the SHG probe.

2. Experimental Procedure and Samples under Study

Experiments were performed for 10 μm thick $(LuBi)_3Fe_5O_{12}$ epitaxial films grown by liquid phase epitaxy on a (111) facet of centrosymmetric GGG substrates. Figure 1a shows the transmission and optical density (OD) spectra of the studied samples, which reveal high absorption at the wavelengths below 600 nm that is typical for garnet films [2]. Namely, the absorption coefficient in the blue spectral range is about four orders of magnitude larger as compared to the wavelength in the near-IR spectral range, which is rather important for the nonlinear optical experiments.

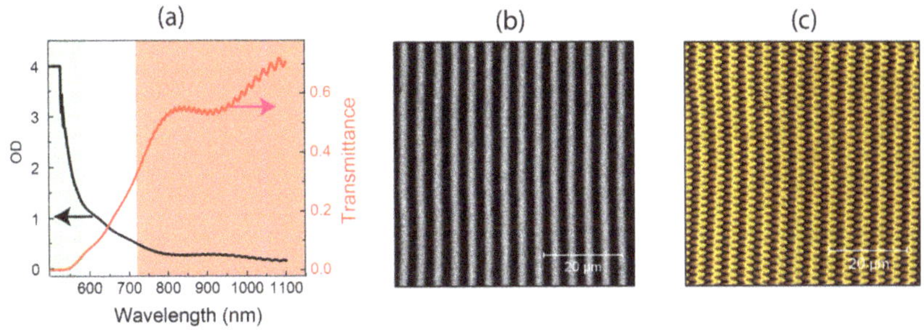

Figure 1. (**a**) Optical density (OD) and transmission spectra of $(LuBi)_3Fe_5O_{12}$ film, high absorption region corresponding to the second and third harmonics wavelengths is in green, high transparency range is in red. (**b**) Optical transmission microscopy image of the film measured for the crossed orientations of the polarizer and the analyzer; and (**c**) MFM image of the $(LuBi)_3Fe_5O_{12}$ film.

Magnetic force microscopy (MFM) experiments of the garnet films were performed using the set-up described in detail in [42]. In brief, we used the atomic force microscope SmartSPM (AIST-NT, now produced by HORIBA Scientific) with a special magnetic tip composed of a multilayered $[Co(2)/Pt(5)]^{10}$ film, the thickness of the corresponding layers in Angstrom is denoted in brackets. Such a composition of the sensitive layer of the tip provided, first, the perpendicular magnetic anisotropy of the film and, second, its small magnetic moment. As a result, the magnetic structure of the studied sample was not disturbed during the MFM measurements.

MFM images of $(LuBi)_3Fe_5O_{12}$ films were obtained in a standard two-pass technique (the so called lift mode), when at the first step the topography of the studied sample is recorded followed by the registration of the magnetic response in the constant distance surface-probe mode at the second step. Our measurements have shown that the films are flat, the root mean square average of the profile height deviations from the mean value (RMS) being about 1 nm. It is worth noting that the MFM probe reveals the distribution of the magnetic stray field along the normal to the surface, while the field associated with the in-plane components of magnetization remain much less pronounced in the MFM profile.

For the nonlinear optical microscopy studies, we used the radiation of an optical parametric oscillator (OPO) system Topol 1050-C (Avesta) with the laser pulse duration of up to 180 fs, wavelength tunable in the spectral range from 740 nm up to 1800 nm, and the mean power of 100 mW. The pump radiation was focused by a Nikon Plan Apo 60× oil objective with the NA = 1.4 in a spot of about 0.7–0.8 μm in diameter, which provided the waist length of the laser beam of a few micrometers, depending on the wavelength of the laser radiation. Mitutoyo M Plan Apo 100x objective with the NA = 0.7 was used for gathering the transmitted radiation at the second or third harmonic's wavelengths, the nonlinear optical signal being detected by the Hamamatsu R4220 photomultiplier (PMT). In our experiments, the $(LuBi)_3Fe_5O_{12}$ film was fixed at the Ratis three-axes piezo scanner stage, providing the positioning accuracy of a few nanometers. Polarizations of the input and harmonics' radiation were controlled by the Glan–Taylor calcite polarizers and the necessary set of color filters placed in front of the PMT were used to block the pump beam passing through the garnet layer.

In accordance with the transmission spectrum of the $(LuBi)_3Fe_5O_{12}$ layer shown in Figure 1a, the studied garnet films are transparent for the fundamental laser radiation in the whole tuning range marked by red in Figure 1a, while second and third harmonics' wavelengths typically fall into the absorption band. This allows the study of the nonlinear and magnetic properties of the interfaces of garnet films with the surrounding media (air and GGG) using the SHG or THG microscopy technique. Besides the evidently high in-plane spatial resolution of the nonlinear optical probe provided by strong focusing of the pump radiation, the in-depth resolution is mostly determined by small escape lengths at harmonics' frequences at the wavelengths shorter than 600 nm. Importantly, the in-depth resolution of the nonlinear microscopy techniques exceeds that of the linear optical one due to the nonlinear dependence of the response on the pump intensity; the higher the order of the nonlinearity, the stronger the spatial localization of the nonlinear optical response at the second or third harmonics' wavelengths.

3. Experimental Results

Figure 1b shows the optical transmission image of the studied $(LuBi)_3Fe_5O_{12}$ film in the absence of the external magnetic field for the crossed polarizations of the incident and transmitted beams. Dark and light regions on the Figure correspond to stripe domains in the bulk of the garnet film with the period of the domain structure of about 4.4 µm; the contrast in the image appears as the result of the opposite in sign Faraday rotation in the adjacent domains with residual magnetization oriented predominantly along the normal to the garnet layer.

A magnetic force microscopy image of the (outer) $(LuBi)_3Fe_5O_{12}$/air interface measured over the 20 µm × 20 µm surface area is shown in Figure 1c. Besides the periodic magnetic structure associated with the aforementioned stripe domains, it reveals a zigzag-like pattern with the period of approximately 1.6 µm associated with the block blind domains. It is worth noting that the high contrast of the MFM signal from the adjacent domains indicate the predominant perpendicular magnetic anisotropy of the film.

Similar conclusions were made based on the studies of the second harmonic generation microscopy of the $(LuBi)_3Fe_5O_{12}$/air interface, as the pump laser beam at the wavelength of 850 nm was incident on the sample from the substrate side (see Figure 2a). In that case, the SHG response at the wavelength of 425 nm is formed by the surface layer of 2–3 µm in thickness (determined by the escape length of the radiation at the SHG wavelength), which allows the study of the surface domain structure of garnet film at this hidden interface.

Figure 2b,d show the SHG intensity patterns for the parallel (panel *b*) and orthogonal (panel *d*) polarizations of the fundamental and SHG beams. In both cases the pump beam was polarized along the stripe domains oriented in the vertical direction on the Figure; the scanning was performed over the surface area of 7 µm × 7 µm. One can see that in the first case, the SHG pattern reveals the stripe domains that appear as the bands with a strong difference in the SHG intensity, while for the crossed polarizations of the pump and SHG waves, a chessboard-like pattern of the SHG intensity is attained. Figure 2c shows the 3D distribution of the SHG intensity from the same surface region as in Figure 2d. It also demonstrates a strongly periodic modulation of the nonlinear optical response with the periods of about 4.2 µm in the horizontal direction and corresponds to the stripe magnetic domains, while the period of 1.6 µm appears along the vertical direction. These images confirm the existence of zigzag-shape closure domains with a predominant in-plane orientation of magnetization, which stems from the symmetry analysis discussed in [42]. Importantly, the stripe domains' alignment and relevant SHG patterns are independent on the crystallographic orientation of the film and follow the direction of the external magnetic field used for their magnetization, which was removed prior to the MFM or nonlinear optical studies.

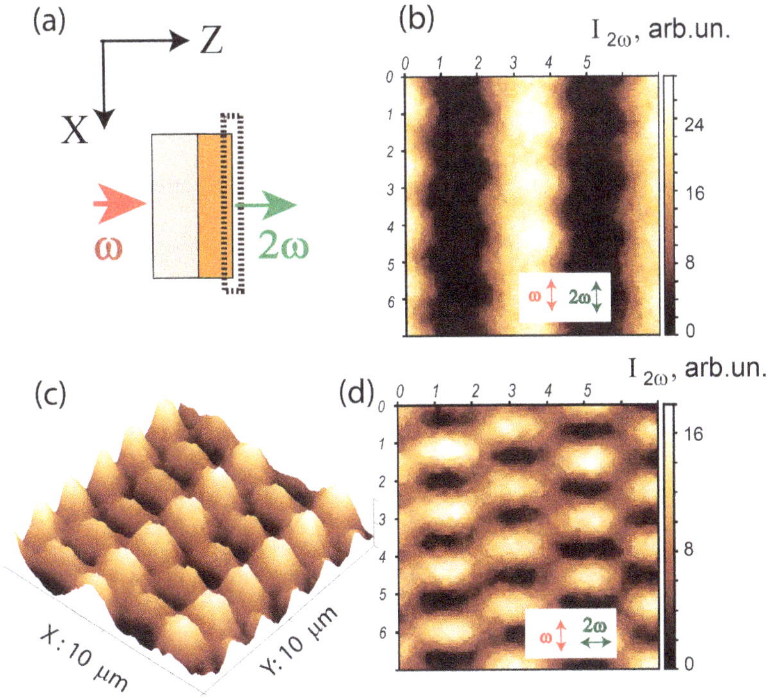

Figure 2. (**a**) Scheme of the light beams propagation in the nonlinear optical microscopy measurements of the $(LuBi)_3Fe_5O_{12}$/air interface; (**b**,**d**) SHG intensity patterns of the $(LuBi)_3Fe_5O_{12}$ film, the polarization plane of the fundamental radiation is oriented in the vertical direction on the figure, the polarization of the generated SHG is vertical (**b**) and horizontal (**d**). (**c**) The 3D SHG intensity pattern corresponding to that shown in panel (**d**).

Similar SHG microscopy experiments were performed for the hidden $(LuBi)_3Fe_5O_{12}$/GGG interface, which is inaccessible by the MFM technique (Figure 3). Really, in that case the MFM probe is separated from the garnet layer by a GGG substrate that is several hundreds of micrometers in thickness. In order to attend the $(LuBi)_3Fe_5O_{12}$/GGG interface by the SHG microscopy probe, we reversed the experimental geometry so that the laser beam was incident on the sample from the film side, as shown schematically in Figure 3a. The effective thickness of the garnet layer participating in the SHG process was estimated by measuring the dependence of the SHG intensity on the coordinate along the normal to the film, z, which is shown in Figure 3b. One can see that the intensity of the SHG response reveals a maximum with the full width at half maximum (FWHM) of about $\approx$2.5 μm situated close to the interface. The coordinate of this SHG maximum z_{max} corresponds to the escape length of the SHG wavelength in the garnet films and thus determines the in-depth resolution of the SHG probe; its position was used as a reper when scanning along the z-direction.

Similarly, we switched to the studies of the third harmonic microscopy of the garnet films. Figure 4 shows the THG intensity patterns for the hidden $(LuBi)_3Fe_5O_{12}$/GGG interface; the experimental scheme is sketched in Figure 4a. The fundamental wavelength in the spectral range of 1200–1600 nm was used, so that the detected THG at 400–530 nm wavelength corresponds to the high absorption of the garnet film. Figure 4b shows that the effective thickness of the layer contributing to the third harmonic generation is about 2 μm, which is less than for the case of the SHG scheme due to the higher-order nonlinearity of the process. Similarly, higher in-plane resolution of the THG probe is attained.

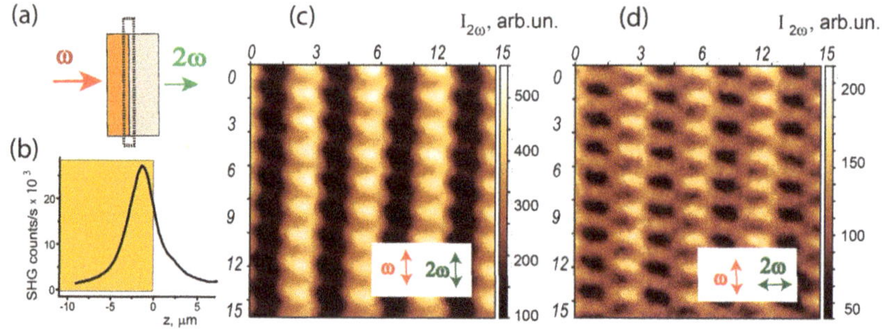

Figure 3. (**a**) Scheme of the light beam propagation in the SHG microscopy measurements of the $(LuBi)_3Fe_5O_{12}$/GGG interface; (**b**) dependence of the SHG intensity on the coordinate within the garnet film, z = 0 corresponds to the garnet/GGG interface. (**c**,**d**) SHG intensity patterns of the $(LuBi)_3Fe_5O_{12}$/GGG interface layer, the polarization plane of the fundamental radiation is oriented in the vertical direction on the figure, the polarization of the generated SHG is vertical (**c**) and horizontal (**d**).

Contrary to the SHG case, the THG intensity pattern for the parallel polarizations of the fundamental and harmonic's beams is constant within the experimental accuracy over the whole scanning range of 10 × 10 μm. This indicates that the relative value of the magnetization-induced components of the THG polarization are negligibly small as compared to the non-magnetic (crystallographic) ones. At the same time, if the detected THG polarization is close to the orthogonal position with respect to the polarization plane of the laser beam and forms the angle of 85° (Figure 4c) or 90° (Figure 4d) with it, pronounced THG patterns appear. One can see a zigzag-type dependence similar to that attained in the SHG response and very much similar to the MFM pattern shown in Figure 1c. For crossed-pump and the THG waves' polarizations, the domain structure demonstrates nearly triangular-like features places in a chess-like order above the stripe domains. If the analyzer is rotated by 5° from the crossed position, an even more detailed magnetic pattern is observed, indicating the spatial resolution of the THG probe of less than 0.5 μm.

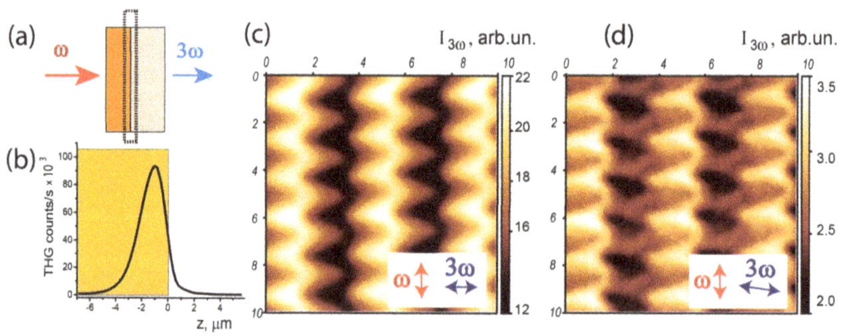

Figure 4. (**a**) Scheme of the light propagation in the THG microscopy measurements; (**b**) the dependence of the averaged THG intensity of the coordinate *z* along the normal to the film. THG intensity patterns measured with the polarization plane of the fundamental radiation oriented in the vertical direction in the figure, the THG polarization is horizontal (**c**) and tilted to about 5° from the horizontal direction (**d**). Fundamental wavelength of 1500 nm was used in these measurements.

4. Discussion

For the case of noncentrosymmetric magnetic media, nonlinear polarization at the second or third harmonics' wavelengths is given by a vector sum of the two types of terms of the electric field at the corresponding wavelengths, the so called *even* and *odd* components with respect to magnetization: $\mathbf{P}_{2\omega,3\omega} \propto \mathbf{E}^{even}_{2\omega,3\omega} + \mathbf{E}^{odd}_{2\omega,3\omega}$. In turn, these field components are determined by the effective *even* and *odd* components of the nonlinear susceptibilities $\chi^{even}_{2\omega,3\omega}(M)=\chi^{even}_{2\omega,3\omega}(-\mathrm{M})$ and $\chi^{odd}_{2\omega,3\omega}(\mathrm{M})=-\chi^{odd}_{2\omega,3\omega}(-\mathrm{M})$, the latter changes its sign under the magnetization reversal. In this simplified description, we restrict ourselves linearly in **M** terms of the magnetization-induced nonlinear optical response at the SHG or THG wavelengths, as they are the strongest in the case of a magnetized sample or in a medium with spontaneous magnetic structure (i.e., in a domain-like one) [35]. Interference of these components of the harmonics' fields provides linear in magnetization modulation of the SHG (THG) intensity, which can be characterized by the corresponding magnetic contrast $\rho_{2\omega,3\omega} \propto \frac{\chi^{odd}_{2\omega,3\omega}}{\chi^{even}_{2\omega,3\omega}}$ [25] and allows the visualization of the magnetic domains with different orientation of the average magnetization. At this stage, magnetization-induced effects in the nonlinear optical response are qualitatively similar to their linear optical analogues, i.e., magnetooptical Kerr and Faraday effects, while the values of the SHG magnetic contrast $\rho_{2\omega}$ is typically one to two orders of magnitude larger.

Evidently, the contrast in the nonlinear optical response from different types of magnetic domains required for their imaging appears only if both *odd* and *even* in **M** effective components of the nonlinear susceptibility exist. The sets of the nonzero components of the nonlinear susceptibility tensors are governed by the crystallographic and magnetic symmetry of a medium. As garnet belongs to the centrosymmetric cubic ($m3m$-symmetric) class, the bulky SHG is suppressed by symmetry arguments and thus the second harmonic (in the electric dipole approximation) should be generated by the interfaces with broken inversion symmetry. At the same time, the bulk SHG effect was registered in substituted garnets with the structure distorted by the presence of large ions such as bismuth or induced by the lattice mismatch at the garnet/GGG interface [43], which seems to be our case. Meanwhile, third harmonic generation is allowed in the bulk of the materials of any symmetry class, including garnets.

If taking into consideration the transparency of garnet films at the fundamental wavelength, transmission coefficients at the second and third harmonic's wavelengths and localization of the pump beam in the laser beam waist, the intensity of the SHG (THG) transmitted through the garnet film can be expressed as

$$I_{m\omega}(\pm M, z, r) = \int_V T_{m\omega}(z)(\chi^{even}_{m\omega} \pm \chi^{odd}_{m\omega}(\pm M))^2 L(z,r)^m dz dr \tag{1}$$

where $m = 2, 3$ corresponds to the number of harmonic (second or third), $L(z,r)$ is the dependence of the fundamental field intensity on the transversal radius-vector **r** and on the coordinate along the direction of the beam propagation, z. Integration over z can be restricted by the waist length according to the expression $L(z,r) = I_0/(1+(\frac{z}{z_R})^2)$, where z_R is the Rayleigh length of the focusing objective and d is the film's thickness, I_0 is the input intensity of the fundamental laser beam, $T_{m\omega}(z) \propto \exp[-\alpha_{m\omega}(d-z)]$ are the intensity transmission functions at the SHG and THG wavelengths, and the integral is taken over the excited volume V of the $(LuBi)_3Fe_5O_{12}$ film. Here we assume that the effective nonlinear susceptibilities $\chi_{m\omega}$ are uniformly distributed in the garnet film and the nonlinear contribution of the GGG substrate is negligibly small, as stems from our experiments. Thus the SHG or THG intensity maxima of the $I_{m\omega}(z)$ dependencies (see Figures 3b and 4b) are attained within the garnet film close to its exit boundary. It is worth noting that the resolution of the THG microscopy technique is higher as compared to the SHG one due to the third-order dependence of its intensity on $L(z,r)$, while it is quadratic for the SHG case.

The analysis of the SHG and THG intensity patterns shown in Figures 3 and 4 is a complicated task, as the epitaxial garnet films are (111) symmetric and the tensors $\hat{\chi}^{(2)}$ and $\hat{\chi}^{(3)}$ contain many nonzero components, both crystallographic (*even*) and magnetization-induced (*odd*) [35,44]. Still, based on the analysis of the SHG intensity patterns measured for the parallel (Figures 2b and 3c) and orthogonal (Figures 2d and 3d) combinations of polarization of the fundamental and SHG beams, one can judge the orientation of the magnetization in domains similarly to those discussed recently in [42] for the garnet/air interface. Namely, the method reveals only the in-plane M components orthogonal to the polarization of the SHG wave, i.e., "vertical" in the first case and "horizontal" in the second. It means that the spontaneous magnetization of stripe domains of the garnet film is tilted with respect to the film's normal, which may be due to the (111) film's symmetry. Similar conclusion on the direction of the magnetization in such films has been proposed based on the analysis of the MFM image of the cleavage (end face) of the garnet film [42]. This provides a large (up to 60%) contrast in the second harmonic intensity generated by the neighbour stripe domains.

Shorter-period modulation of the SHG intensity in the direction along the stripe domains observed at the interfaces of the $(LuBi)_3Fe_5O_{12}$ layer with air (Figure 2) and with the GGG substrate (Figure 3) is associated with the nonlinear response of the block (closure) domains; the modulation is better seen for the crossed polarizations of the SHG and pump radiation (panels *d* in Figures 2 and 3). In that case, the maximum contrast of the SHG intensity is attained for the adjacent closure domains placed along the horizontal line in Figures 2d or 3d, which correspond to the opposite signs of the "vertical" **M** component. If taking into account the analogous analysis of the SHG intensity pattern for the parallel polarizations of the SHG and pump beams, we have to conclude that the in-plane magnetization of the closure domains forms a nonzero angle with the magnetic stripes. This is probably the reason for the formation of the zigzag-shape patterns attained both in the MFM measurements (see Figure 1c) and in the THG response (Figure 4).

Based on the results of the experiments described in this paper, we may conclude that the organization of magnetic domains at the hidden $(LuBi)_3Fe_5O_{12}$/GGG interface is quite similar to that of the $(LuBi)_3Fe_5O_{12}$/air interface; even the periods of stripe and block domains nearly coincide. This shows that a discontinuity of the garnet crystalline structure plays the most important role in the formation of the closure surface magnetic domains, while the influence of the surrounding media, namely crystalline gallium gadolinium garnet or air in our case, is less important.

5. Conclusions

Summing up, we performed experimental studies of the magnetic domain structure of 10 μm thick *Bi*, *Lu*-substituted epitaxial garnet films by a combination of magnetic force microscopy and nonlinear optical microscopy techniques based on the effects of the second and third harmonics generation. In the latter case, strong localization of the pump radiation of a pulsed femtosecond laser source along with the nonlinear power dependence of the registered harmonics' signal allowed to the visualization of the magnetic stripe domains in the bulk of the garnet film through a pronounced contrast in the average second and third harmonics' intensity in adjacent domains. Based on the symmetry analysis of the nonlinear optical response, we conclude that magnetization of this type of domain is tilted with respect to the film's normal as well as to the stripe domains .

Moreover, we applied the nonlinear optical technique to approach and investigate the magnetic organization of the buried interface of the $(LuBi)_3Fe_5O_{12}$ epitaxial layer with the gallium gadolinium garnet substrate, which is inaccessible for the MFM studies. We found that a zigzag-shape interface domain appear above the bulk stripe ones, similarly to the upper $(LuBi)_3Fe_5O_{12}$/air interface studied quite recently. This confirms that the main role is played here by the discontinuity of the garnet crystalline structure for the magnetic domains' organization, while the type of neighbour medium is less important.

We also present the first experimental demonstration of the efficiency of the the third harmonic generation microscopy for the study of magnetic domains in garnet films. In spite of relatively small linear values in magnetization contribution to the third harmonic response, the THG microscopy for the nearly crossed polarizations of the fundamental and THG beams allows the visualization of the surface magnetic closure domains at the hidden garnet/GGG interface and provides even higher spatial resolution of the magnetic structure as compared to magnetization-induced second harmonic generation.

Author Contributions: Conceptualization, A.T., A.M. and T.M.; methodology, M.T. and A.M.; software, A.T. and A.M.; validation, M.T., T.M. and A.M.; formal analysis, A.T. and A.M.; investigation, M.T., A.M. and A.T.; resources, A.M.; data curation, M.T. and A.M.; writing—original draft preparation, T.M. and A.T.; writing—review and editing, A.T. and A.M.; visualization, A.T.; supervision, T.M.; project administration, A.T. and M.T.; funding acquisition, T.M. All authors have read and agreed to the published version of the manuscript.

Funding: This research was funded by the Russian Science Foundation, Grant 19-72-20103-P. MFM studies were supported by Russian Science Foundation, Grant 19-19-00607-P.

Institutional Review Board Statement: Not applicable.

Data Availability Statement: The data presented in this paper are available upon request.

Acknowledgments: The authors acknowledge useful discussions with M. Logunov.

Conflicts of Interest: The authors declare no conflict of interest.

References

1. Hubert , A.; Schäfer, R. *Magnetic Domains. The Analysis of Magnetic Microstructures*; Springer: Berlin/Heidelberg, Germany, 1998.
2. Zvezdin, A.K.; Kotov, V.A. *Modern Magnetooptics and Magnetooptical Materials*; CRC Press: Boca Raton, FL, USA, 1997; 404p.
3. Hashimoto, R.; Itaya, T.; Uchida, H.; Funaki, Y.; Fukuchi, S. Properties of Magnetic Garnet Films for Flexible Magneto-Optical Indicators Fabricated by Spin-Coating Method. *Materials* **2022**, *15*, 1241. [CrossRef]
4. Arsad, A.Z.; Zuhdi, A.W.M.; Ibrahim, N.B.; Hannan, M.A. Recent Advances in Yttrium Iron Garnet Films: Methodologies, Characterization, Properties, Applications, and Bibliometric Analysis for Future Research Directions. *Appl. Sci.* **2023**, *13*, 1218. [CrossRef]
5. Baños-López, E.; Sánchez-De Jesús, F.; Cortés-Escobedo, C.A.; Barba-Pingarrón, A.; Bolarín-Miró, A.M. Enhancement in Curie Temperature of Yttrium Iron Garnet by Doping with Neodymium. *Materials* **2018**, *11*, 1652. [CrossRef]
6. Shimokawa, K.; Dohnomae, H.; Mukai, T.; Yamada, H.; Matsuda, H.; Daimon, M. Nanocrystalline garnet disks for magneto-optical recording media. *J. Magn. Magn. Mater.* **1996**, *154*, 271. [CrossRef]
7. Costa, J.D.; Figeys, B.; Sun, X.; Van Hoovels, N.; Tilmans, H.A.; Ciubotaru, F.; Adelmann, C. Compact tunable YIG-based RF resonators. *Appl. Phys. Lett.* **2021**, *118*, 162406. [CrossRef]
8. Aichele, T.; Lorenz, A.; Hergt, R.; Görnert, P. Garnet layers prepared by liquid phase epitaxy for microwave and magneto-optical applications—A review. *Cryst. Res. Technol. J. Exp. Ind. Crystallogr.* **2003**, *38*, 575. [CrossRef]
9. Nikitov, S.; Filimonov, Y.; Vysotsky, S.; Khivintsev, Y.; Pavlov, E. Yttrium iron garnet based phononic-magnonic crystal. In Proceedings of the IEEE International Ultrasonics Symposium, Dresden, Germany, 7–10 October 2012; pp. 1240–1243.
10. Heinz, B.; Bracher, T.; Schneider, M.; Wang, Q.; Laagel, B.; Friedel, A.M.; Breitbach, D.; Steinert, S.; Meyer, T.; Kewenig, M.; et al. Chumak, Propagation of Spin-Wave Packets in Individual Nanosized Yttrium Iron Garnet Magnonic Conduits. *Nano Lett.* **2020**, *20*, 4220. [CrossRef]
11. Chumak, A.V.; Vasyuchka, V.I.; Serga, A.A.; Hillebrands, B. Magnon spintronics. *Nat. Phys.* **2015**, *11*, 453. [CrossRef]
12. Randoshkin, V.V.; Chervonenkis, A.Y. *Applied Magneto-Optics* Énergoatomizdat: Moscow, Russia, 1990.
13. Vlasko-Vlasov, V.K.; Dedukh, l.M.; Nikitenko, V.I. Domain structure of yttrium iron garnet single crystals. *Sov. Phys. JETP* **1976**, *44*, 1208.
14. McCord, J. Progress in magnetic domain observation by advanced magneto-optical microscopy. *J. Phys. D Appl. Phys.* **2015**, *48*, 333001. [CrossRef]
15. Vértesy, G. Coercive Properties of Magnetic Garnet Films. *Crystals* **2023**, *13*, 946. [CrossRef]
16. Yu, X.; DeGrave, J.P.; Hara, Y.; Hara, T.; Jin, S.; Tokura, Y. Observation of the magnetic skyrmion lattice in a MnSi nanowire by Lorentz TEM. *Nano Lett.* **2013**, *14*, 3755. [CrossRef]
17. Zhou, T.; Cherukara, M.; Phatak, C. Differential programming enabled functional imaging with Lorentz transmission electron microscopy. *NPJ Comput. Mater.* **2021**, *7*, 141. [CrossRef]
18. Kazakova, O.; Puttock, R.; Barton, C.; Corte-León, H.; Jaafar, M.; Neu, V.; Asenjo, A. Frontiers of magnetic force microscopy. *J. Appl. Phys.* **2019**, *125*, 060901. [CrossRef]
19. Temiryazev, A.G.; Tikhomirova, M.P.; Fedorov, I. Surface domains in inhomogeneous yttrium iron garnet films. *J. Magn. Magn. Mater.* **2003**, *258*, 580. [CrossRef]
20. Xia, W.X.; Chun, Y.S.; Aizawa, S.; Yanagisawa, K.; Krishnan, K.M.; Shindo, D.; Tonomura, A. Investigation of magnetic structure and magnetization process of yttrium iron garnet films by Lorentz microscopy and electron holography. *J. Appl. Phys.* **2010**, *108*, 123919. [CrossRef]

21. Da Silva, N.R.; Möller, M.; Feist, A.; Ulrichs, H.; Ropers, C.; Schäfer, S. Nanoscale Mapping of Ultrafast Magnetization Dynamics with Femtosecond Lorentz Microscopy. *Phys. Rev. X* **2018**, *8*, 031052.
22. Vysokikh, Y.E.; Shelaev, A.V.; Prokopov, A.R.; Shevyakov, V.I.; Yu, S. Krasnoborodko, Magnetic domain structure investigation of Bi: YIG-thin films by combination of AFM and cantilever-based aperture SNOM. *J. Phys. Conf. Ser.* **2016**, *741*, 012190. [CrossRef]
23. Shen, R. *The Principles of Nonlinear Optics*; Wiley: New York, NY, USA, 1984.
24. Akhmediev, N.N.; Borisov, S.B.; Zvezdin, A.K.; Lubchanskii, I.L.; Melikhov, Y.V. Nonlinear optical susceptibility of magnetically-ordered crystals. *Fiz. Tverd. Tela* **1985**, *27*, 1075.
25. Pan, R.P.; Wei, H.D.; Shen, Y.R. Optical second harmonic generation from magnetized surfaces. *Phys. Rev. B* **1989**, *39*, 1229. [CrossRef]
26. Reif, J.; Zink, J.C.; Schneider, C.-M.; Kirschner, J. Effects of surface magnetism on optical second harmonic generation. *Phys. Rev. Lett.* **1991**, *67*, 2878. [CrossRef]
27. Vollmer, R.; Jin, Q.Y.; Regensburger, H.; Kirschner, J. Second harmonic generation from magnetic surfaces and thin films. *J. Magn. Magn. Mater.* **1999**, *198*, 611. [CrossRef]
28. Spierings, G.; Koutsos, V.; Wierenga, H.A.; Prins, M.W.J.; Abraham, D.; Rasing, T. Interface magnetism studied by optical second harmonic generation. *J. Magn. Magn. Mater.* **1993**, *121*, 109. [CrossRef]
29. Mcgilp, J.F.; Carroll, L.; Fleischer, K.; Cunniffe, J.P.; Ryan, S. Magnetic second-harmonic generation from interfaces and nanostructures. *JMMM* **2010**, *322*, 1488. [CrossRef]
30. Kolmychek, I.A.; Radovskaya, V.V.; Mamonov, E.A.; Maydykovskiy, A.I.; Sadovnikov, A.V.; Sheshukova, S.E.; Nikitov, S.A.; Temiryazeva, M.P.; Gusev, N.S.; Fraerman, A.A.; et al. Interface-induced optical effects in magnetic two- and three-layer films. *JMMM* **2021**, *528*, 167780. [CrossRef]
31. Murzina, T.V.; Kim, E.M.; Kapra, R.V.; Aktsipetrov, O.A.; Kravets, A.F.; Inoue, M.; Kuznetsova, S.V.; Ivanchenko, M.V.; Lifshits, V.G. Magnetization-induced optical third-harmonic generation in Co and Fe nanostructures. *Phys. Rev. B* **2006**, *73*, 140404.
32. Radovskaya, V.V.; Mamonov, E.A.; Pashen'kin, I.Y.; Gusev, N.S.; Murzina, T.V. Magneto-optical effects in second harmonic generation from W/Co/Pt nanofilms. *J. Phys. Conf. Ser.* **2021**, *2015*, 012117. [CrossRef]
33. Kirilyuk, V.; Kirilyuk, A.; Rasing, T. A combined nonlinear and linear magneto-optical microscopy. *Appl. Phys. Lett.* **1997**, *70*, 2306. [CrossRef]
34. Lyubchanskii, I.L. Theory of optical second harmonic generation in magnetic films with domain structure. *J. Magn. Magn. Mater.* **1996**, *157*, 307. [CrossRef]
35. Petukhov, A.V.; Lyubchanskii, I.L.; Rasing, T. Theory of nonlinear magneto-optical imaging of magnetic domains and domain walls. *Phys. Rev. B* **1997**, *56*, 2680. [CrossRef]
36. Squier, J.A.; Müller, M.; Brakenhoff, G.J.; Wilson, K.R. Third harmonic generation microscopy. *Opt. Express* **1998**, *3*, 22. [CrossRef]
37. Borile, G.; Sandrin, D.; Filippi, A.; Anderson, K.I.; Romanato, F. Label-Free Multiphoton Microscopy: Much More Than Fancy Images. *Int. J. Mol. Sci.* **2021**, *22*, 2657. [CrossRef]
38. Gavgiotaki, E.; Filippidis, G.; Tsafas, V.; Bovasianos, S.; Kenanakis, G.; Georgoulias, V.; Tzardi, M.; Agelaki, S.; Athanassakis, I. Third Harmonic Generation microscopy distinguishes malignant cell grade in human breast tissue biopsies. *Sci. Rep.* **2020**, *10*, 11055. [CrossRef]
39. Ogata, Y.; Vorobyev, A.; Guo, C. Optical Third Harmonic Generation Using Nickel Nanostructure-Covered Microcube Structures. *Materials* **2018**, *11*, 501. [CrossRef]
40. Raghunathan, V.; Deka, J.; Menon, S.; Biswas, R.; As, L.K. Nonlinear Optics in Dielectric Guided-Mode Resonant Structures and Resonant Metasurfaces. *Micromachines* **2020**, *11*, 449. [CrossRef]
41. Okhlopkov, K.I.; Zilli, A.; Tognazzi, A.; Rocco, D.; Fagiani, L.; Mafakheri, E.; Bollani, M.; Finazzi, M.; Celebrano, M.; Shcherbakov, M.R.; et al. Tailoring Third-Harmonic Diffraction Efficiency by Hybrid Modes in High-Q Metasurfaces. *Nano Lett.* **2021**, *21*, 10438. [CrossRef]
42. Temiryazeva, M.; Mamonov, E.; Maydykovskiy, A.; Temiryazev, A.; Murzina, T. Magnetic Domain Structure of $Lu_{2.1}Bi_{0.9}Fe_5O_{12}$ Epitaxial Films Studied by Magnetic Force Microscopy and Optical Second Harmonic Generation. *Magnetochemistry* **2022**, *8*, 180. [CrossRef]
43. Kumar, P.; Maydykovskiy, A.I.; Levy, M.; Dubrovina, N.V.; Aktsipetrov, O.A. Second harmonic generation study of internallygenerated strain in bismuth-substituted iron garnet films. *Opt. Express* **2010**, *18*, 1076. [CrossRef]
44. Birrs, R.R. *Symmetry and Magnetizm*; North-Holland Publishing Co.: Amsterdam, The Netherlands, 1966.

applied sciences

MDPI

Article

Femtosecond Optical Nonlinearity of Nanodiamond Suspensions

Gennady M. Mikheev [1,*], Viatcheslav V. Vanyukov [2], Tatyana N. Mogileva [1], Konstantin G. Mikheev [1], Alexander N. Aleksandrovich [1], Nicholas A. Nunn [3,4] and Olga A. Shenderova [3]

1 Udmurt Federal Research Center, Institute of Mechanics, Ural Branch of the Russian Academy of Sciences, 426067 Izhevsk, Russia; mogileva@udman.ru (T.N.M.); k.mikheev@udman.ru (K.G.M.); alexan@udman.ru (A.N.A.)
2 Department of Physics and Mathematics, Institute of Photonics, University of Eastern Finland, 80101 Joensuu, Finland; viatcheslav.vanyukov@uef.fi
3 Adamas Nanotechnologies, Inc., Raleigh, NC 27617, USA; nnunn@adamasnano.com (N.A.N.); oshenderova@adamasnano.com (O.A.S.)
4 Department of Chemistry, North Carolina State University, Raleigh, NC 27695, USA
* Correspondence: mikheev@udman.ru; Tel.: +7-34-1221-8955

Abstract: High pressure-high temperature (HP-HT) nanodiamonds and detonation nanodiamonds have unique optical properties and are promising materials for various applications in photonics. In this work, for the first time, comparative studies of the nonlinear optical properties of aqueous suspensions of HP-HT and detonation nanodiamonds under femtosecond laser excitation are performed. Using the z-scan technique, it was found that for the same laser pulse parameters HP-HT nanodiamonds exhibited optical limiting due to two-photon absorption while detonation nanodiamonds exhibited saturable absorption accompanied by short-term optical bleaching, revealing the different electronic-gap structures of the two types of nanodiamonds. The saturable absorption properties of detonation nanodiamonds are characterized by determining the saturable and non-saturable absorption coefficients, the saturation intensity, and the ratio of saturable to non-saturable losses. The nonlinear absorption in HP-HT nanodiamonds is described with the nonlinear absorption coefficient that decreases with decreasing concentration of nanoparticles linearly. The results obtained show the possibility of using aqueous suspensions of nanodiamonds for saturable absorption and optical limiting applications.

Keywords: nanodiamonds; nonlinear optics; nonlinear refraction; nonlinear absorption; two-photon absorption; saturable absorption; optical limiting

Citation: Mikheev, G.M.; Vanyukov, V.V.; Mogileva, T.N.; Mikheev, K.G.; Aleksandrovich, A.N.; Nunn, N.A.; Shenderova, O.A. Femtosecond Optical Nonlinearity of Nanodiamond Suspensions. *Appl. Sci.* **2021**, *11*, 5455. https://doi.org/10.3390/app11125455

Academic Editor: Alessandro Belardini

Received: 14 May 2021
Accepted: 10 June 2021
Published: 11 June 2021

Publisher's Note: MDPI stays neutral with regard to jurisdictional claims in published maps and institutional affiliations.

1. Introduction

From the invention of lasers to the present time, the intensive study of the nonlinear optical properties of various materials has continued. The rapid development of nanotechnology in recent years has led to a wide variety of nanomaterials, including nanocarbons, which have nonlinear optical properties and that are of great interest for various applications. To date, there are a large number of works devoted to the study of the nonlinear optical properties of carbon nanotubes, graphene, and various composites based on them. These works are aimed at investigating, e.g., optical limiting (a nonlinear decrease of the transmitted light intensity) [1] and saturable absorption (a nonlinear increase of the transmitted light intensity) [2] to develop and create optical limiters of nanosecond laser pulses [3–13] and passive laser shutters for generating picosecond and femtosecond laser pulses [14–24], respectively. Nanodiamond, being another form of nanocarbon material, is also attractive due to its unique properties [25–33]. Among the various methods for producing nanodiamonds [27,34], the most prevalent are the detonation synthesis method [35] and the method of grinding diamonds synthesized at high pressure and high temperature (HP-HT) [36].

The nonlinear optical properties of detonation nanodiamonds and various complexes based on them have been extensively studied. Similar to carbon nanotubes and graphene, detonation nanodiamonds have been studied for the development and creation of nonlinear optical filters operating in a wide wavelength range to protect the eyes and sensitive optical components from the damaging effect of powerful laser pulses of nanosecond duration. It was found that in addition to nonlinear absorption, the nonlinear scattering of light has a significant effect on the nonlinear decrease in the optical transmittance with increasing incident light fluence in the nanosecond time domain [37–41]. Nonlinear absorption and nonlinear scattering have been found, depending on the nanodiamond particles size and concentration in suspensions [42–44]. It was also found that the functionalization of detonation nanodiamonds with other nanoparticles and complexes suitable for this purpose leads to a further decrease in the nonlinear transmittance coefficient [41,45–48]. It has been also shown that aqueous suspensions of detonation nanodiamonds exhibit a saturable absorption (a short-term increase in the transmittance) under nanosecond [49] and femtosecond [50] excitations. Despite the number of studies of nonlinear optical properties of detonation nanodiamonds, to the best of our knowledge, there is only one work on the nonlinear optical properties of HP-HT nanodiamonds. In that work, the observation of optical limiting in an aqueous suspension of HP-HT nanodiamonds under excitation by nanosecond laser pulses at a wavelength of 532 nm was reported [51]. This is of interest for further investigation of the nonlinear optical properties of HP-HT nanodiamonds by exploring the femtosecond time domain and to compare the nonlinear optical response of those with the detonation nanodiamonds under the same experimental conditions.

In this work, we show that aqueous suspensions of HP-HT nanodiamonds exhibit nonlinear optical properties in the femtosecond time domain. Moreover, we show that aqueous suspensions of detonation and HP-HT nanodiamonds with an average size of ~10 nm possess different nonlinear optical properties under femtosecond laser excitation. Specifically, suspensions of HP-HT nanodiamonds exhibit optical limiting originating from the two-photon absorption, while suspensions of detonation nanodiamonds demonstrate optical self-bleaching due to saturable absorption.

2. Materials and Methods

HP-HT nanodiamonds were obtained from Van Moppes, Geneva, Switzerland. The HP-HT nanodiamond particles are monocrystals with a substitutional nitrogen content of ~100 ppm. The purification of nanodiamonds from carbon was carried out by processing in the air at a temperature of 500 °C for two hours, followed by treatment in hydrochloric acid at a temperature of 70 °C for two hours. The nanoparticles were dispersed in deionized water at 1.65 wt% concentration and the suspensions were stored in plastic vessels.

Detonation nanodiamonds were synthesized by an explosion of an oxygen-deficient explosive mixture of trinitrotoluene and hexogen at a weight ratio of 1:1 in a closed steel chamber. The resulting detonation soot contains the nanodiamond particles (more than 30%), the other allotropic states of carbon, and various metallic impurities. Purification of the diamond soot in the mixture at the high temperature reduces the amount of impurities down to 1 wt%. These detonation nanodiamonds received from a vendor were additionally purified at Adamas Nanotechnologies with HCl, reducing the metal content to 0.4 wt%. Detonation synthesis leads to the formation of nanodiamonds with a primary particle size of about 5 nm. The natural agglomeration of nanodiamonds leads to the formation of solid particles of much larger sizes. To obtain nanoparticles with a given average size the purified nanodiamonds were suspended in deionized water by sonication and processed in a planetary mill for 4 h using zirconia beads. Additional sp^2-carbon from milling was removed by treatment at 400 °C in the air for 3 h. After this high-temperature treatment, the resulting product was resuspended in deionized water at 1% w/v by sonication. Centrifugation at 25,000× g forces was used to extract nanodiamond particles with an average diameter of ~10 nm. The nitrogen content in the detonation nanodiamonds is ~10,000 ppm.

The major differences in HP-HT and detonation nanodiamonds are in the size of the primary particles, the presence of aggregates consisting of few primary particles in detonation nanodiamonds and nitrogen content. While treatments to reduce sp^2-carbon content were performed, residual sp^2-carbon is still present in both types of nanodiamond particles, with a higher content expected in detonation nanodiamonds due to the smaller size of primary particles.

For surface and particle size analysis, small droplets of fabricated suspensions of HP-HT and detonation nanodiamonds were placed on a flat glass surface and dried to form a uniform film. The topography images of the nanodiamond films were acquired with an NTEGRA Probe Nanolaboratory atomic-force microscope (AFM) in semicontact mode. The particle size distribution was attained by processing the atomic-force microscope images with Image Analysis 3.5.0 software (NT-MDT Spectrum Instruments, Zelenograd, Russia) based on the combination of the correlation analysis and the method of sections at a certain relative height. The AFM images and the particle size distribution of the nanodiamond films are shown in Figure 1. The performed measurements reveal that more than 75% of the HP-HT and detonation nanodiamonds have sizes in the ranges of 9–20 and 6–13 nm, respectively.

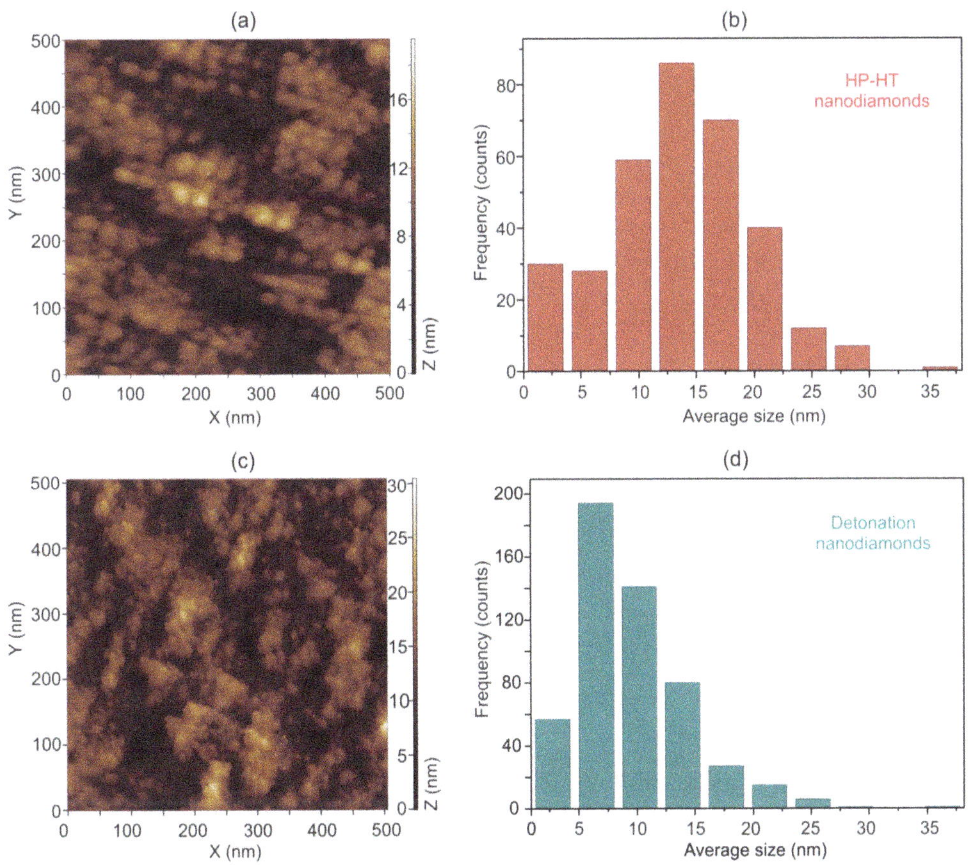

Figure 1. Atomic force microscopy images of (**a**) HP-HT and (**c**) detonation nanodiamond films. The image processing reveals the HP-HT and detonation nanodiamond particle size distributions shown in figures (**b**) and (**d**), correspondingly.

Raman spectra of the attained nanodiamond films were recorded using Horiba LabRam HR800 Raman spectrometer (Lille, France) with a 632.8 nm laser. A 100× objective with the attenuation of the incident radiation (neutral filter D1) was used to provide an intensity (less than about 5 kW/cm^2) below the graphitization of nanodiamond material at 632.8 nm [29,52]. The measured Raman spectra of detonation and HP-HT nanodiamond films show pronounced peaks with frequency shifts of 1329 and 1330 cm^{-1}, respectively (see Figure 2a). Taking into account the known dependences of the Raman shifts of carbon nanomaterials on the excitation wavelength [53] and the nanoparticle size [54], one can attribute these peaks to nanodiamonds (sp^3-form of carbon). The decomposition of the spectra presented in Figure 2a shows that the full width at half maximum (FWHM) of diamond lines of HP-HT and detonation nanodiamonds are 8 and 22 cm^{-1}, respectively. It is noteworthy that along with the specified peaks, the Raman spectra exhibit broad luminescence bands caused by defects of various natures contained in HP-HT and detonation nanodiamonds [29,55].

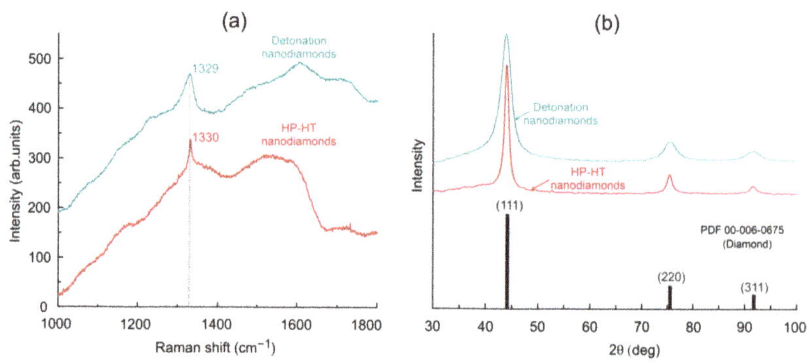

Figure 2. (**a**) Raman spectra of HP-HT and detonation nanodiamonds film and (**b**) X-ray diffraction patterns of HP-HT and detonation nanodiamond powders. The X-ray diffraction pattern of diamond (Powder Diffraction File 00-006-0675) is presented for a reference.

For the X-ray diffraction studies, nanodiamond powders were obtained by drying the suspensions at room temperature and subsequent mechanical grinding. When obtaining diffraction patterns, the studied powders were placed on a glass substrate. The diffraction pattern of the substrate without powder was recorded separately. To obtain the diffraction pattern of the studied powder, the corresponding subtraction of the substrate response from the experimental data was performed.

The crystalline structure of the nanodiamonds was studied using Bruker D2 PHASER X-ray diffractometer with a copper-based X-ray tube generating radiation at a wavelength of 0.1541 nm ($K_{\alpha 1}$ line). The diffraction patterns of the detonation and HP-HT nanodiamonds powders (see Figure 2b) contain three solitary peaks in the range of 2θ variation from 25° to 100°. These peaks are observed at 2θ angles of 43.9°, 75.3°, and 91.5°. They correspond to X-ray diffraction on the (111), (220), (311) planes of diamond crystallites with interplanar spacings of d_{111} = 0.2060 nm, d_{220} = 0.1261 nm, and d_{311} = 0.1075 nm, respectively, where the subscripts at d denote crystallographic planes. Note that according to the reference data of diamond (see Figure 2b), the intensities of the diffraction lines of monochromatic radiation on the (111), (220), (311) planes are equal to 100, 25, and 16 arb. units, respectively (see Powder Diffraction File (PDF) 00-006-0675). Figure 2b shows that for both diffraction patterns, the ratio between peak amplitudes at angles 2θ equal to 43.9°, 75.3°, and 91.5° is the same. Processing the measured X-ray diffraction pattern using TOPAS 4.2 software revealed that the mean crystallite sizes (LVol-IB) of the HP-HT and detonation nanodiamond particles are 8.6 and 2.7 nm, respectively. Thus, the measurements on an X-ray diffractometer show that all the samples under study consist of diamond material.

For the optical measurements, suspensions of both types of nanodiamonds were prepared at three concentrations. Figure 3 shows the optical densities of aqueous suspensions of HP-HT and detonation nanodiamonds at three different concentrations measured in a 1 mm thick optical quartz cell. The measurements were performed using a PerkinElmer Lambda 650 (Shelton, WA, USA) two-beam spectrophotometer relative to the same 1 mm-thick quartz cell filled with distilled water. It can be seen that for the suspensions of both types of nanodiamonds, the optical density decreases monotonically with an increase of light wavelength which is the characteristic of nanodiamonds [29,32,49,56,57]. For the suspensions of HP-HT nanodiamonds, the increase in optical density with decreasing wavelength is more pronounced than for the suspensions of detonation nanodiamonds. This can be explained by stronger light scattering in the suspension of HP-HT nanodiamonds caused by the following reasons: (i) the slightly larger average nanoparticle size of the HP-HT nanodiamond suspension compared to the average nanoparticle size of the detonation nanodiamond suspension; (ii) the larger "effective" refractive index of HP-HT nanodiamonds compared to the "effective" refractive index of detonation nanodiamonds due to the difference in their crystal structures. For both types of nanodiamonds the optical density measured at different wavelengths is proportional to the concentration (see insets in Figure 3a,b).

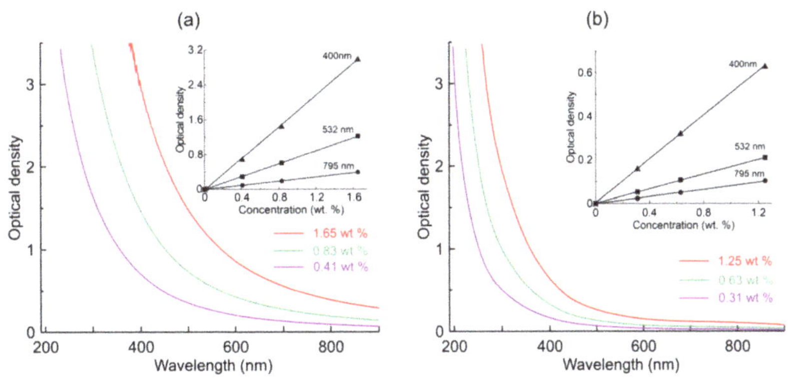

Figure 3. The optical density of the (**a**) HP-HT and (**b**) detonation nanodiamond suspensions in 1 mm thick quartz cell. Insets show the corresponding optical density as a function of concentration measured at three different wavelengths of 400, 532, and 795 nm.

Studies of the nonlinear optical properties of nanodiamond suspensions were carried out using the z-scan technique [58] with femtosecond linearly polarized laser pump (wavelength 795 nm, pulse duration 120 fs, and the repetition rate of 1 kHz). In the experiments, simultaneous measurements of transmittance in the closed- and open-aperture configurations were performed (Figure 4). The laser radiation was focused on a cell with a focal length of 75 mm. The cell had 1 mm-thick quartz walls with a 1 mm opening distance. The laser beam waist radius was w_0 = 12.9 μm, which corresponds to the Rayleigh length of z_0 = 0.65 mm ($z_0 = \pi w_0^2/\lambda$, where λ is the wavelength). The scanning of the cell was performed along the optical z-axis of the focused laser beam using computer numerical control with a step of 0.1 mm with multiple averaging of the pulse energies E_{oa} and E_{ca} recorded using "open-aperture" and "closed-aperture" photodetectors, respectively (see Figure 4).

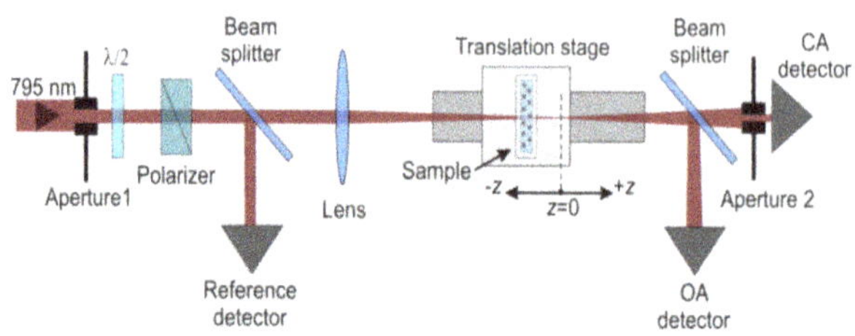

Figure 4. Sketch of the z-scan experimental setup.

In the experiments, the dependences of the nonlinear transmittance coefficients $T_{oa}(z) = E_{oa}(z)/E_{in}$ (with an open-aperture) and $T_{ca}(z) = E_{ca}(z)/E_{in}$ (with a closed-aperture) were measured during scanning of the cell along the z-axis, where E_{in} is the incident laser pulse energy measured using a reference photodetector. This made it possible to determine the dependences of the normalized nonlinear transmittance coefficients $T_{oa,n}(z) = T_{oa}(z)/T_{oa}(z = \infty)$ and $T_{ca,n}(z) = T_{ca}(z)/T_{ca}(z =\infty)$ with open- and closed-apertures respectively, where $T_{oa}(z = \infty)$ and $T_{ca}(z = \infty)$ are the optical transmittance coefficients at open- and closed-apertures, respectively, obtained far from the beam waist, i.e., far from z = 0. The energy of laser pulses incident on the cell was varied using a polarizer and a half-wave plate. To visualize nonlinear refraction in the experiments, the ratio of nonlinear transmittance obtained in a closed- aperture to the data obtained in an open-aperture $t_{nr}(z) = T_{ca,n}(z)/T_{oa,n}(z)$ was determined [59].

3. Results

The dependences $T_{oa,n}(z)$ and $t_{nr}(z)$ obtained for a cell filled with distilled water (a,b) and an empty cell (c,d) at E_{in} = 200 nJ are presented in Figure 5a–d. The dependence $T_{oa,n}(z)$ obtained for a cell with distilled water shows a small dip in the vicinity of $z/z_0 = 0$ indicating a weak nonlinear absorption. At the same time, the experimental data $T_{oa,n}(z)$ obtained for an empty cell (Figure 5c) shows no corresponding dip. This means that the observed weak nonlinear absorption for a cell with distilled water occurs because of distilled water and not because of the cell. The nonlinear absorption in distilled water originates from multi-photon absorption [60]. Comparison of the closed/open-aperture data for a cell filled with water (Figure 5b) and an empty cell (Figure 5d) shows that the observed nonlinear refraction leading to self-focusing of the femtosecond laser beam occurs mainly in the quartz walls of the optical cell. The observed cluster of dots at the vicinity of $z/z_0 = 0$ in Figure 5d arises from the quartz walls-air-quartz walls boundary and not affecting on the amplitude of the normalized transmittance. Closed/open-aperture measurements were also performed for both the studied types of nanodiamonds with the results shown in Figure 5e,f). Adding the nanoparticles to water has no noticeable impact on the nonlinear refraction of the suspensions as one can observe by comparing Figure 5b,d–f) indicating weak nonlinear refraction in both suspensions for given concentrations.

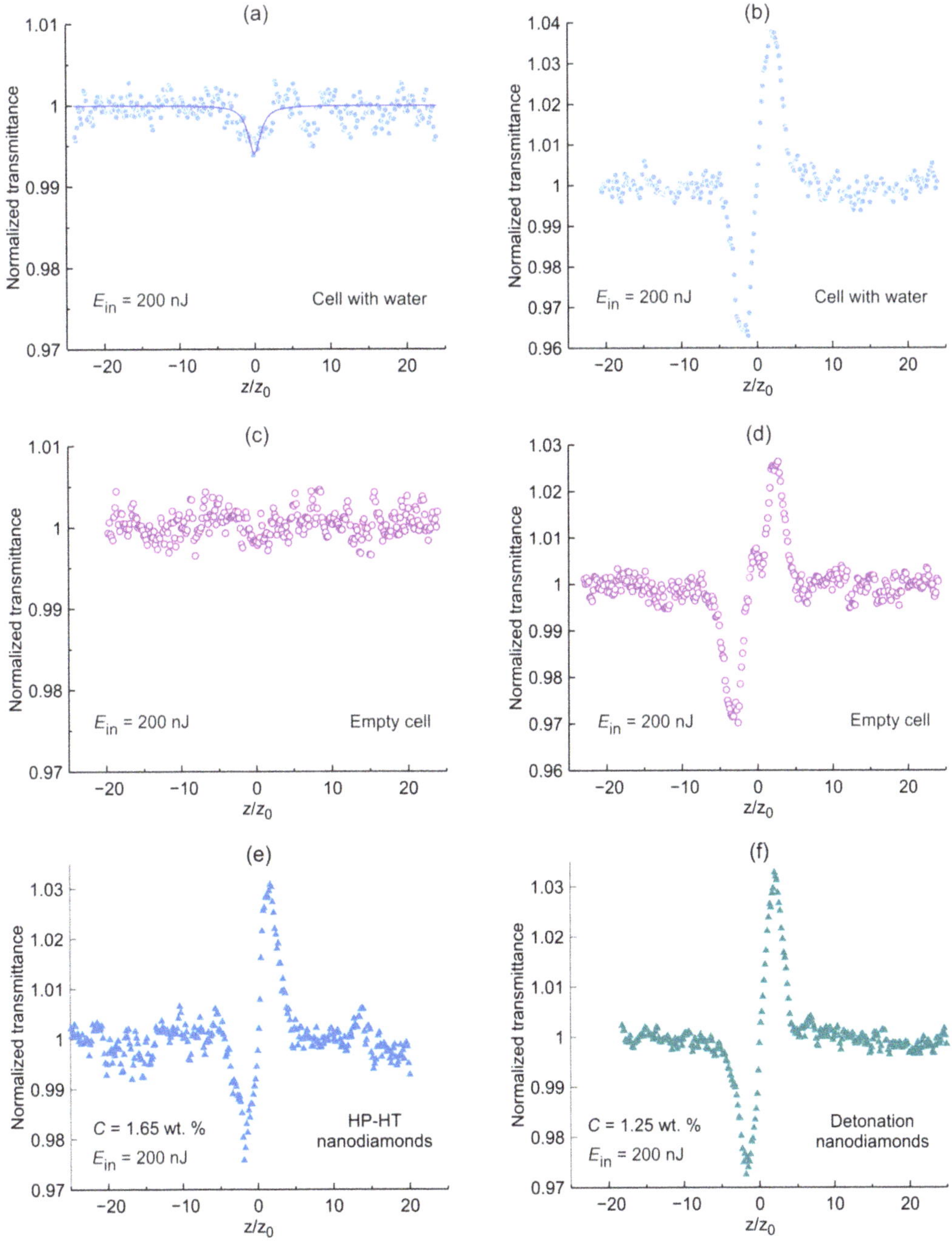

Figure 5. The normalized transmittance of the quartz cell with (**a**) distilled water and (**c**) an empty cell obtained in the open-aperture z-scan. The ratio of the nonlinear transmittance obtained in the closed to the data obtained in the open-aperture z-scan for the quartz cell (**b**) with distilled water, (**e**) with HP-HT nanodiamond suspension, (**f**) with detonation nanodiamond suspension, and (**d**) an empty cell. The measurements were performed at the incident laser pulse energy of 200 nJ.

The results of the open-aperture normalized transmittance measurements for aqueous suspensions of HP-HT nanodiamonds are shown in Figure 6. The measurements were performed at the same incident pulse energy of E_{in} = 200 nJ. In order to reveal the dependence of the nonlinear optical properties, i.e., nonlinear absorption on the nanoparticle concentration C, we performed the measurements for the C ranging from 0.1 to 1 wt%. It is shown that the higher the concentration, the stronger the nonlinear absorption, the lower the transmittance at z/z_0. Even at a low concentration of 0.08 wt%, the nonlinear transmittance exhibits a visible dip. It is noteworthy that the dependencies shown in Figure 6 are symmetric relative to z/z_0 = 0 indicating prominent optical limiting properties of femtosecond laser pulses.

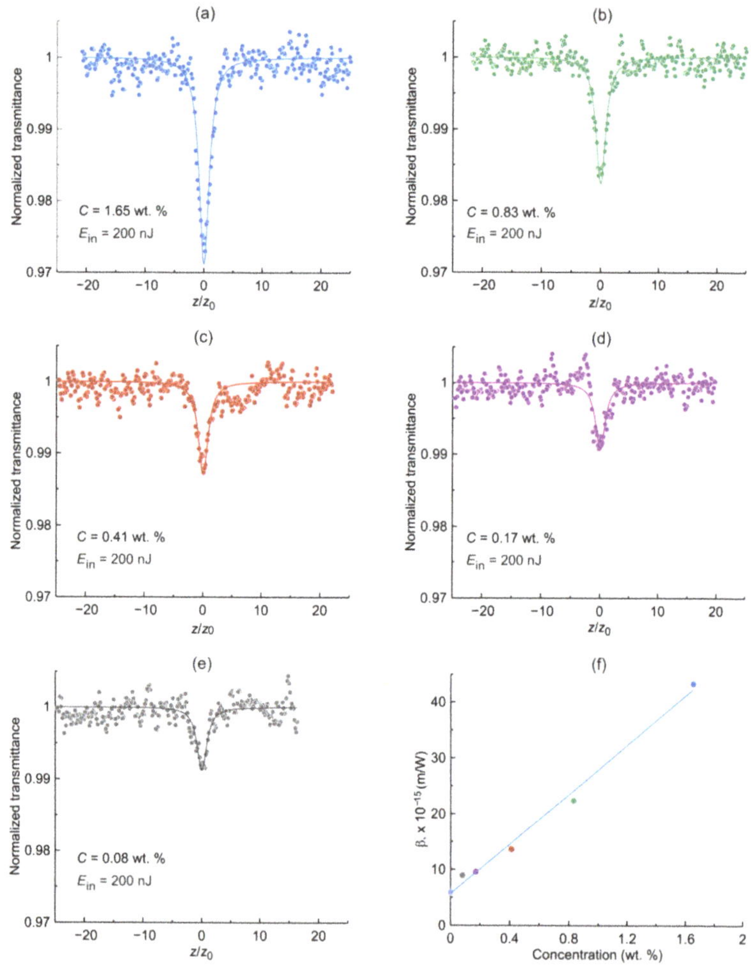

Figure 6. (**a**–**e**) Normalized transmittance of suspensions of HP-HT nanodiamonds with different concentrations recorded in the open-aperture z-scan. Dots correspond to the experimental data while solid lines represent the result of the fitting with Equation (2). Figure (**f**) represents the dependence of the calculated two-photon absorption coefficient β as a function of concentration C.

One can assume that the mechanism of optical limiting in suspensions of HP-HT nanodiamonds is two-photon absorption, in which the absorption coefficient α_{TPA} of the medium depends on the radiation intensity I according to the equation [61]

$$\alpha_{TPA}(I) = \alpha + \beta I, \tag{1}$$

where α and β are the coefficients of linear and nonlinear absorption, respectively. This can be verified by approximating the experimental data presented in Figure 6 with the well-known equation (see [61]) derived to describe two-photon absorption for the open-aperture z-scan for a Gaussian laser beam and temporally Gaussian pulse:

$$T_{oa,n} = 1 + \sum_{m=1}^{\infty} [-q(z, t=0)]^m / (m+1)^{3/2}, \tag{2}$$

where $q_0(z,\ t=0) = \beta I_0(t=0) L_{eff} / (1 + z^2/z_0^2), L_{eff} = \left(1 - e^{-\alpha L}\right)/\alpha\,,\ \alpha = -lnT_0/L,\ I_0$ $(t=0) = E_{in}/[\pi w_0^2 \Delta t \frac{\sqrt{\pi}}{2\sqrt{ln2}}]$ (see [58]), $I_0(t)$ is the on-axis irradiance at focus (i.e., z = 0), Δt is the pulse duration (FWHM) of a Gaussian laser pulse, T_0 is the linear transmittance of the studied sample with thickness L. Equation (2) with the first four terms from an infinite series of elements is in good agreement with the experimental z-scan data shown in Figure 6. The values of β found as a result of the approximation of the experimental data, depending on the concentration of nanoparticles C, are shown in Figure 6f and can be expressed as β(m/W) = (5.7 + 22.1 $\times$ C(wt%)) $\times$ 10^{-15}. It follows from the approximation, that the defined β does not vanish when the concentration of nanoparticles in suspension is tending to zero confirming the earlier obtained z-scan data for a cell filled with distilled water (see Figure 5a). Based on the above, one can conclude that the two-photon absorption well describes the mechanism of the nonlinear behavior in aqueous suspensions of HP-HT nanodiamonds.

The results of the normalized transmittance recorded in the open-aperture z-scan for a fixed concentration of aqueous suspension of detonation nanodiamonds are shown in Figure 7. We have not presented the measurements for the various concentrations as for HP-HT nanodiamonds because those have been already reported [49]. Instead, measurements of the normalized transmittance with variation of the incident laser pulse energy have been performed. In contrast to HP-HT nanodiamonds, suspensions of detonation nanodiamonds exhibit an increase in the nonlinear transmittance with increasing incident light intensity. In the range of incident laser pulse energies, the experimental curves of $T_{oa,n}(z)$ presented in Figure 7 are symmetrical relative to the focal point z/z_0 = 0. This suggests that the observed phenomenon is not associated with laser-induced irreversible bleaching [62,63], but originates from the saturable absorption which has been extensively studied in graphene, carbon nanotubes [15,23,24,64,65] and has also been reported in detonation nanodiamonds with larger particle sizes under femtosecond excitation [50]. Similar to graphene and carbon nanotubes the observed saturable absorption in detonation nanodiamonds is characterized by a short-term decrease of its absorption coefficient during the passing of a light pulse of high intensity and caused by the electronic transition between two energy levels.

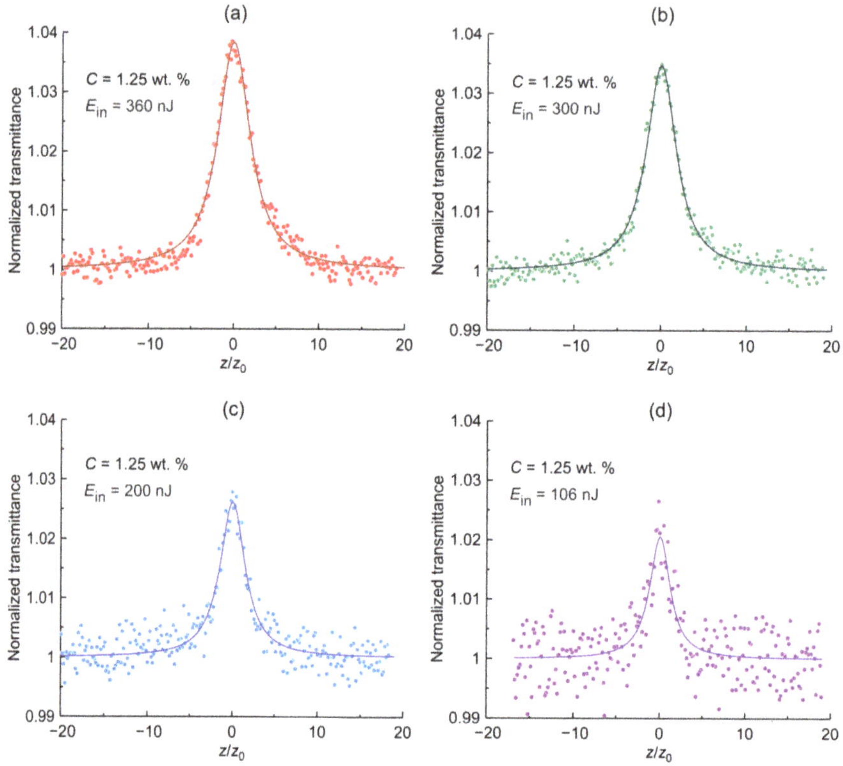

Figure 7. (**a–d**) Normalized transmittance of suspensions of detonation nanodiamonds obtained in the open-aperture z-scan for a fixed concentration of 1.25 wt% for various incident laser pulse energies ranging from 106 to 360 nJ. Dots represent the experimental data while solid lines represent the result of the fitting with Equation (5).

In accordance with [65,66], the absorption coefficient at saturable absorption is given by:

$$\alpha_{SA}(I) = \alpha_{ns} + \alpha_0/(1 + I/I_{sat}) \quad (3)$$

where α_{ns} is the linear absorption coefficient due to the presence of light losses in the medium that are not associated with saturable absorption, α_0 is the coefficient characterizing the absorption in a two-level energy system, where the population moves under the action of resonant laser pumping, leading to equalization of the level populations at high intensities (saturation of absorption) [67], I_{sat} is the saturation intensity, i.e., the intensity at which α_0 is halved. It should be noted that the linear absorption coefficient $\alpha_{SA}(I = 0) = \alpha_{ns} + \alpha_0$, and at $I \gg I_{sat}$ $\alpha_{SA}(I \to \infty) = \alpha_{ns}$, is true, i.e., even at very high pump intensities, complete bleaching of the medium does not occur. The coefficients α_0 and α_{ns} are commonly referred to as saturable and nonsaturable absorptions, respectively.

As was established above, the observed multiphoton absorption in pure distilled water is weak and therefore one can neglect its effect on the saturable absorption. In order to find the dependence of nonlinear transmittance on the incident intensity of a suspension of nanoparticles exhibiting saturable absorption the following differential equation should be solved:

$$dI/dz\prime = -\alpha_{SA} I, \quad (4)$$

which is describing the propagation of the light pulse in a nonlinear medium [61] with the absorption coefficient presented in the form (3). For this case, the exact solution of Equation (4) yields to [49]:

$$T_{oa,n} = \left[\frac{I_{out}/I_{sat} + 1 + \alpha_0/\alpha_{ns}}{I_{in}/I_{sat} + 1 + \alpha_0/\alpha_{ns}} \right]^{-\alpha_0/\alpha_{ns}}, \tag{5}$$

where I_{out} and I_{in} are the output and incident radiation intensities. Saturable absorption leads to a slight increase in the intensity at the output of the cell (see Figure 7), therefore, the following expression is valid $I_{out} = T_0 \times I_{in}$. For a Gaussian beam during z-scan, the relation $I_{in} = I_0/(1 + (z/z_0)^2)$ is satisfied. Taking all this into account, one can arrive at the following equation describing the dependence of the normalized transmittance on the sample position:

$$T_{oa,n} = \left\{ \frac{T_0 \times A + (1+B) \times \left(1 + (z/z_0)^2\right)}{A + (1+B) \times \left(1 + (z/z_0)^2\right)} \right\}^{-B}, \tag{6}$$

where $A = I_0/I_{sat}$ and $B = \alpha_0/\alpha_{ns}$.

Equation (6) makes it possible to approximate the experimental dependences shown in Figure 7 with two parameters A and B. Considering that $lnT_0 = -(\alpha_{ns} + \alpha_0) \times L$, where T_0 is the linear transmittance of the suspension, we arrive at the $I_{sat} = 110\ \mathrm{GW/cm^2}$, $\alpha_{ns} = 2.0\ \mathrm{cm^{-1}}$ and $\alpha_0 = 0.36\ \mathrm{cm^{-1}}$. It is seen from Figure 7 that Equation (6) with the defined parameters approximates the experimental data well. It should be noted that the obtained values of α_{ns} and α_0 are defined for the nanoparticle concentration of C = 1.25 wt% and depend linearly on concentration [49], i.e., $\alpha(\mathrm{cm^{-1}}) = 1.6 \times$ C (wt%), α_0 $(\mathrm{cm^{-1}}) = 0.24 \times$ C (wt%). This means that for both the HP-HT and detonation nanodiamonds, the nonlinear absorption coefficients, i.e., two-photon absorption β and saturable absorption α_0 coefficients, linearly depend on the nanoparticles concentration. The obtained α_{ns} and α_0 are similar to those obtained for the single-digit detonation nanodiamonds [49]. The attained saturation intensity I_{sat} which is concentration independent [49] is the same order as for the aqueous suspensions of detonation nanodiamonds with an average particle size of 50 nm under the femtosecond pulse excitation [50] but 11×10^3 times higher than that for the single-digit detonation nanodiamonds under the nanosecond pulse excitation [49]. A similar difference in saturation intensities in nanosecond and femtosecond time domains has been observed in a graphene polymer composite [68].

Another important parameter characterizing the saturable absorption properties and the applicability of the materials for the passive mode-locking is the ratio of the saturable to non-saturable losses [65]:

$$R = \{\exp(-\alpha_{ns}L) - \exp[-(\alpha_{ns} + \alpha_0)L]\}/[1 - \exp(-\alpha_{ns}L)], \tag{7}$$

which should be as high as possible. Since the products $\alpha_{ns}L$ and $\alpha_0 L$ are small, we arrive at $R = \alpha_0/\alpha_{ns} = 0.18$ which is lower than for graphene and single-walled carbon nanotubes $(R \approx 1)$ used for passive mode-locking and Q-switching [65].

It should be added that in our experiments, despite the nonlinear absorption, the nonlinear refraction in the nanodiamond suspension due to heat absorption does not manifest itself. This is explained by the fact that the nonlinear absorption observed in the experiments is weak. For example, in the HP-HT nanodiamond suspension at a nanoparticle concentration of C = 1.65 wt% at a laser pulse energy of 200 nJ, the nonlinear absorption is only about 3% (see Figure 6a), and in the suspension of detonation nanodiamonds at the concentration of nanoparticles C = 1.25 wt% at the energy of laser pulses 360 nJ the nonlinear absorption does not exceed 4%.

4. Discussion

Comparing the obtained results of the nonlinear optical properties of HP-HT and detonation nanodiamond suspensions in the femtosecond time domain one can conclude that they have similar nonlinear refraction but different nonlinear absorption properties. The nonlinear refraction is weak for both types of nanodiamonds and has no impact on the nonlinear refraction of the aqueous suspension with a nanoparticle concentration of about 1 wt%. Under the same experimental conditions, two-photon absorption occurs in suspensions of HP-HT nanodiamonds leading to a decrease in the transmittance, while saturable absorption appears in an aqueous suspension of detonation nanodiamonds leading to self-bleaching indicating that the electronic structures of HP-HT and detonation nanodiamonds differ from each other significantly. It is well known that the energy of the bandgap E_g of a pure bulk diamond is 5.4 eV and as a result there is no optical absorption in such material. The photon energy of the femtosecond laser we used is 1.56 eV, which is about 3.5 times less than E_g. This means that one-photon or two-photon absorptions, as well as saturable absorption at a wavelength of 795 nm in pure bulk diamond, are impossible. However, as can be seen from Figure 3, aqueous suspensions of HP-HT and detonation nanodiamonds are semitransparent in the optical range and their optical densities increase monotonically with decreasing wavelength. In accordance with [56], this is caused by light scattering (the cross-section of which decreases with an increase in the wavelength of incident radiation) and optical absorption that occurs in detonation nanodiamonds in the energy range of 1–2 eV. In a recent paper [69], it was shown that the characteristic extinction spectrum (see Figure 3) of detonation nanodiamonds with a size less than 100 nm is mainly determined by absorption rather than a scattering of light. The saturable absorption we observed in a suspension of detonation nanodiamonds at a photon energy of 1.56 eV is in agreement with these studies. The absence of saturable absorption in a suspension of HP-HT nanodiamonds indicates that there are no energy transitions that effectively absorb radiation at a wavelength of 795 nm. On the other hand, the observation of nonlinear absorption described by two-photon absorption shows that the suspensions of HP-HT nanodiamonds have an absorption band with transition energy of about 3.1 eV, which agrees with the optical density spectrum shown in Figure 3. The significant difference in the femtosecond nonlinear behavior of two types of nanodiamonds may be caused by their different structure.

Indeed, as established above, the average crystallite sizes of HP-HT and detonation nanodiamonds are 8.6 and 2.7 nm, respectively. Taking into account the sizes of the nanoparticles studied (see Figure 1), this means that HP-HT nanodiamonds are predominantly composed of one or two single crystallites, while detonation nanodiamond particles are conglomerates predominantly composed of two to five single crystallites. The difference in the structures of nanoparticles of the two types of nanodiamonds is also indicated by the difference in their Raman spectra presented in Figure 2a. The diamond Raman lines of the HP-HT and detonation nanodiamond spectra with frequency shifts of 1330 and 1329 cm^{-1}, respectively, differ from each other in line width and shape. The FWHM of the diamond line of detonation nanodiamonds is approximately 2.8 times greater than the corresponding value found for HP-HT nanodiamonds. In addition, the diamond line of detonation nanodiamonds is asymmetric with a more pronounced broadening towards lower wavenumbers. All these findings are in agreement with the literature data, and, according to Refs. [34] and [70], indicate a greater number of defects present in detonation nanodiamonds. Overall, monocrystalline HP-HT nanodiamonds have a uniform structure with a concentration of lattice defects (such as dislocations or twinning) lower than in detonation nanodiamonds containing a diamond core and various lattice defects, transient sp^3/sp^2 layer, and sp^2 surface shell that may carry various surface functional groups [70]. The sp^2-carbon content in detonation nanodiamonds is expected to be higher than in HP-HT nanodiamond due to the smaller size of primary particles. It should be noted that along with the presence of the sp^2-carbon [71], the appearance of energy transitions lower than E_g in nanodiamond can be caused by the presence of diamond-like shells [72], the dimer chains on the surface of diamond nanoparticles [56], as well as the numerous defects, vacancies, and impurities (see, for example, [29,73–75]). Independently of the origin of the two-photon absorption observed in an aqueous suspension of HP-HT nanodiamonds, it can be used for the optical limiting to protect sensitive optical components and eyes from the damaging effect of high-power femtosecond laser radiation with a wavelength of about 800 nm.

5. Conclusions

The nonlinear absorption of suspensions of HP-HT and detonation nanodiamonds at a wavelength of 795 nm of femtosecond excitation differ from each other significantly. At the same parameters of femtosecond laser pumping and experimental conditions, saturable absorption occurred in an aqueous suspension of detonation nanodiamonds accompanied by short-term nonlinear bleaching, while in an aqueous suspension of HP-HT nanodiamonds, two-photon absorption appeared, leading to optical limiting. The nonlinear absorption coefficient of the aqueous suspension of HP-HT nanodiamonds decreased linearly with the decreasing nanoparticle concentration. The important parameters characterizing the saturable absorption of detonation nanodiamonds in an aqueous suspension were determined. The results obtained show that the energy structures of HP-HT and detonation nanodiamonds differ from each other significantly. Aqueous suspensions of HP-HT nanodiamonds can be used for the optical limiting of high-power femtosecond laser pulses while suspensions of detonation nanodiamonds possess properties of the saturable absorber and can be employed for shortening the duration and to remove the pedestal ("tail") of the laser pulses.

Author Contributions: Conceptualization, G.M.M. and V.V.V.; methodology, G.M.M., V.V.V. and K.G.M.; software, V.V.V.; validation, G.M.M. and O.A.S.; formal analysis, V.V.V.; investigation, G.M.M., V.V.V., K.G.M., T.N.M. and A.N.A.; resources, N.A.N. and O.A.S.; data curation, T.N.M.; writing—original draft preparation, G.M.M. and V.V.V.; writing—review and editing, G.M.M., V.V.V., N.A.N. and O.A.S.; visualization, T.N.M.; supervision, G.M.M. and O.A.S.; project administration, G.M.M.; funding acquisition, G.M.M. All authors have read and agreed to the published version of the manuscript.

Funding: This research was funded by the Ministry of Education and Science of the Russian Federation, state registration number AAAA-A19-119021890083-0, by the Academy of Finland grants no. 298298 and 323053, Flagship Programme "Photonics Research and Innovation (PREIN)" decision no. 320165.

Institutional Review Board Statement: Not applicable.

Informed Consent Statement: Not applicable.

Data Availability Statement: The data presented in this study are available on request from the corresponding author.

Acknowledgments: The authors wish to thank S.G. Bystrov for his help in experiments. This study was performed using equipment of the Shared Use Center "Center of Physical and Physicochemical Methods of Analysis and Study of the Properties and Surface Characteristics of Nanostructures, Materials, and Products" UdmFRC UB RAS.

Conflicts of Interest: The authors declare no conflict of interest.

References

1. Van Stryland, E.W.; Wu, Y.Y.; Hagan, D.J.; Soileau, M.J.; Mansour, K. Optical limiting with semiconductors. *J. Opt. Soc. Am. B* **1988**, *5*, 1980–1988. [CrossRef]
2. Haus, H.A. Theory of mode locking with a fast saturable absorber. *J. Appl. Phys.* **1975**, *46*, 3049–3058. [CrossRef]
3. Vivien, L.; Lanc, P.; Riehl, D.; Hache, F.; Anglaret, E. Carbon nanotubes for optical limiting. *Carbon* **2002**, *40*, 1789–1797. [CrossRef]
4. Izard, N.; Billaud, P.; Riehl, D.; Anglaret, E. Influence of structure on the optical limiting properties of nanotubes. *Opt. Lett.* **2005**, *30*, 1509–1511. [CrossRef] [PubMed]
5. Kumar, R.; Kumar, A.; Verma, N.; Philip, R.; Sahoo, B. FeCoCr alloy-nanoparticle embedded bamboo-type carbon nanotubes for non-linear optical limiting application. *J. Alloys Compd.* **2020**, *849*, 156665. [CrossRef]

6. Wang, J.; Chen, Y.; Blau, W.J. Carbon nanotubes and nanotube composites for nonlinear optical devices. *J. Mater. Chem.* **2009**, *19*, 7425–7443. [CrossRef]
7. Wang, J.; Fruchtl, D.; Sun, Z.; Coleman, J.N.; Blau, W.J. Control of optical limiting of carbon nanotube dispersions by changing solvent parameters. *J. Phys. Chem. C* **2010**, *114*, 6148–6156. [CrossRef]
8. Mikheev, G.M.; Mogileva, T.N.; Okotrub, A.V.; Bulatov, D.L.; Vanyukov, V.V. Nonlinear light scattering in a carbon nanotube suspension. *Quantum Electron.* **2010**, *40*, 45–50. [CrossRef]
9. Lim, G.-K.; Chen, Z.-L.; Clark, J.; Goh, R.G.S.; Ng, W.-H.; Tan, H.-W.; Friend, R.H.; Ho, P.K.H.; Chua, L.-L. Giant broadband nonlinear optical absorption response in dispersed graphene single sheets. *Nat. Photonics* **2011**, *5*, 554–560. [CrossRef]
10. Tereshchenko, S.A.; Podgaetskii, V.M.; Gerasimenko, A.Y.; Savel'ev, M.S. Threshold effect under nonlinear limitation of the intensity of high-power light. *Quantum Electron.* **2015**, *45*, 315–320. [CrossRef]
11. Sousani, A.; Motiei, H.; Najafimoghadam, P.; Hasanzade, R. Synthesis of nanocomposites based on carbon nanotube/smart copolymer with nonlinear optical properties. *Opt. Mater.* **2017**, *67*, 172–179. [CrossRef]
12. Krivenkov, R.Y.; Mogileva, T.N.; Mikheev, K.G.; Okotrub, A.V.; Mikheev, G.M. Heat-induced dip of optical limiting threshold in carbon nanotube aqueous suspension. *J. Phys. Chem. C* **2018**, *122*, 16339–16345. [CrossRef]
13. Savelyev, M.S.; Gerasimenko, A.Y.; Podgaetskii, V.M.; Tereshchenko, S.A.; Selishchev, S.V.; Tolbin, A.Y. Conjugates of thermally stable phthalocyanine J-type dimers with single-walled carbon nanotubes for enhanced optical limiting applications. *Opt. Laser Technol.* **2019**, *117*, 272–279. [CrossRef]
14. Sakakibara, Y.; Tatsuura, S.; Kataura, H.; Tokumoto, M.; Achiba, Y. Near-infrared saturable absorption of single-wall carbon nanotubes prepared by laser ablation method. *Jpn. J. Appl. Phys.* **2003**, *42*, L494–L496. [CrossRef]
15. Yamashita, S.; Inoue, Y.; Maruyama, S.; Murakami, Y.; Yaguchi, H.; Jablonski, M.; Set, S.Y. Saturable absorbers incorporating carbon nanotubes directly synthesized onto substrates and fibers and their application to mode-locked fiber lasers. *Opt. Lett.* **2004**, *29*, 1581–1583. [CrossRef] [PubMed]
16. Pushkin, A.V.; Migal, E.A.; Tokita, S.; Korostelin, Y.V.; Potemkin, F.V. Femtosecond graphene mode-locked Fe:ZnSe laser at 4.4 μm. *Opt. Lett.* **2020**, *45*, 738–741. [CrossRef]
17. Hasan, T.; Sun, Z.; Wang, F.; Bonaccorso, F.; Tan, P.H.; Rozhin, A.G.; Ferrari, A.C. Nanotube-polymer composites for ultrafast photonics. *Adv. Mater.* **2009**, *21*, 3874–3899. [CrossRef]
18. Chen, H.-R.; Wang, Y.-G.; Tsai, C.-Y.; Lin, K.-H.; Chang, T.-Y.; Tang, J.; Hsieh, W.-F. High-power, passively mode-locked Nd:GdVO4 laser using single-walled carbon nanotubes as saturable absorber. *Opt. Lett.* **2011**, *36*, 1284–1286. [CrossRef]
19. Xu, J.-L.; Li, X.-L.; Wu, Y.-Z.; Hao, X.-P.; He, J.-L.; Yang, K.-J. Graphene saturable absorber mirror for ultra-fast-pulse solid-state laser. *Opt. Lett.* **2011**, *36*, 1948–1950. [CrossRef]
20. Martinez, A.; Sun, Z. Nanotube and graphene saturable absorbers for fibre lasers. *Nat. Photonics* **2013**, *7*, 842–845. [CrossRef]
21. Chu, H.; Zhao, S.; Li, T.; Yang, K.; Li, G.; Li, D.; Zhao, J.; Qiao, W. Dual-wavelength passively Q-Switched Nd, Mg: LiTaO3 laser with a monolayer graphene as saturable absorber. *IEEE J. Sel. Top. Quantum Electron.* **2015**, *21*, 343–347. [CrossRef]
22. Zhang, S.; Shen, C.; Kislyakov, I.; Dong, N.; Ryzhov, A.A.; Zhang, X.; Belousova, I.M.; Nunzi, J.M.; Wang, J. A photonic crystal based broadband graphene saturable absorber. *Opt. Lett.* **2018**, *44*, 4785–4788. [CrossRef]
23. Wang, Y.; Zhang, B.; Yang, H.; Hou, J.; Su, X.; Sun, Z.; He, J. Passively mode-locked solid-state laser with absorption tunable graphene saturable absorber mirror. *J. Light. Technol.* **2019**, *37*, 2927–2931. [CrossRef]
24. Vanyukov, V.V.; Mikheev, K.G.; Nasibulin, A.G.; Svirko, Y.; Mikheev, G.M. Enhanced saturable absorption in the laser-treated free-standing carbon nanotube films. *Opt. Lett.* **2020**, *45*, 5377–5380. [CrossRef]
25. Shumilov, F.A.; Levitskii, V.S. Immobilization of single particles of detonation nanodiamonds in Langmuir-Blodgett films using octadecylamine. *J. Superhard Mater.* **2019**, *41*, 412–420. [CrossRef]
26. Shvidchenko, A.V.; Eidelman, E.D.; Vul, A.Y.; Kuznetsov, N.M.; Stolyarova, D.Y.; Belousov, S.I.; Chvalun, S.N. Colloids of detonation nanodiamond particles for advanced applications. *Adv. Colloid Interface Sci.* **2019**, *268*, 64–81. [CrossRef]
27. Basso, L.; Cazzanelli, M.; Orlandi, M.; Miotello, A. Nanodiamonds: Synthesis and application in sensing, catalysis, and the possible connection with some processes occurring in space. *Appl. Sci.* **2020**, *10*, 4094. [CrossRef]
28. Shang, C.; Liu, T.; Zhang, F.; Chen, F. Nonlinear optical response of gold-decorated nanodiamond hybrids. *Compos. Commun.* **2020**, *19*, 74–81. [CrossRef]
29. Mikheev, K.G.; Mogileva, T.N.; Fateev, A.E.; Nunn, N.A.; Shenderova, O.A.; Mikheev, G.M. Low-power laser graphitization of high pressure—High temperature nanodiamond films. *Appl. Sci.* **2020**, *10*, 3329. [CrossRef]
30. Vervald, A.M.; Plastinin, I.V.; Burikov, S.A.; Dolenko, T.A. Fluorescence of nanodiamonds under the influence of surfactants. *Fuller. Nanotub. Carbon Nanostructures* **2020**, *28*, 83–89. [CrossRef]
31. Katsiev, K.; Solovyeva, V.; Mahfouz, R.; Abou-Hamad, E.; Peng, W.; Idriss, H.; Kirmani, A.R. Fresh insights into detonation nanodiamond aggregation: An X-ray photoelectron spectroscopy, thermogravimetric analysis, and nuclear magnetic resonance study. *Eng. Rep.* **2021**. [CrossRef]
32. Berdichevskiy, G.M.; Vasina, L.V.; Ageev, S.V.; Meshcheriakov, A.A.; Galkin, M.A.; Ishmukhametov, R.R.; Nashchekin, A.V.; Kirilenko, D.A.; Petrov, A.V.; Martynova, S.D.; et al. A comprehensive study of biocompatibility of detonation nanodiamonds. *J. Mol. Liq.* **2021**, *332*, 115763. [CrossRef]

33. Pedroza-Montero, F.; Santacruz-Gómez, K.; Acosta-Elías, M.; Silva-Campa, E.; Meza-Figueroa, D.; Soto-Puebla, D.; Castaneda, B.; Urrutia-Bañuelos, E.; Álvarez-Bajo, O.; Navarro-Espinoza, S. Thermometric characterization of fluorescent nanodiamonds suitable for biomedical applications. *Appl. Sci.* **2021**, *11*, 4065. [CrossRef]
34. Mochalin, V.N.; Shenderova, O.; Ho, D.; Gogotsi, Y. The properties and applications of nanodiamonds. *Nat. Nanotechnol.* **2012**, *7*, 11–23. [CrossRef]
35. Dolmatov, V.Y. Detonation-synthesis nanodiamonds: Synthesis, structure, properties and applications. *Russ. Chem. Rev.* **2007**, *76*, 339–360. [CrossRef]
36. Shenderova, O.A.; McGuire, G.E. Science and engineering of nanodiamond particle surfaces for biological applications. *Biointerphases* **2015**, *10*, 030802. [CrossRef] [PubMed]
37. Koudoumas, E.; Kokkinaki, O.; Konstantaki, M.; Couris, S.; Korovin, S.; Detkov, P.; Kuznetsov, V.; Pimenov, S.; Pustovoi, V. Onion-like carbon and diamond nanoparticles for optical limiting. *Chem. Phys. Lett.* **2002**, *357*, 336–340. [CrossRef]
38. Mikheev, G.M.; Puzyr, A.P.; Vanyukov, V.V.; Purtov, K.V.; Mogileva, T.N.; Bondar, V.S. Nonlinear scattering of light in nanodiamond hydrosol. *Tech. Phys. Lett.* **2010**, *36*, 358–361. [CrossRef]
39. Vanyukov, V.V.; Mikheev, G.M.; Mogileva, T.N.; Puzyr, A.P.; Bondar, V.S.; Svirko, Y.P. Polarization-sensitive nonlinear light scattering and optical limiting in aqueous suspension of detonation nanodiamonds. *J. Opt. Soc. Am. B* **2014**, *31*, 2990–2995. [CrossRef]
40. Papagiannouli, I.; Bourlinos, A.B.; Couris, S. Nonlinear optical properties of colloidal carbon nanoparticles: Nanodiamonds and carbon dots. *RSC Adv.* **2014**, *4*, 40152–40160. [CrossRef]
41. Muller, O.; Pichot, V.; Merlat, L.; Spitzer, D. Optical limiting properties of surface functionalized nanodiamonds probed by the Z-scan method. *Sci. Rep.* **2019**, *9*, 519. [CrossRef] [PubMed]
42. Josset, S.; Muller, O.; Schmidlin, L.; Pichot, V.; Spitzer, D. Nonlinear optical properties of detonation nanodiamond in the near infrared: Effects of concentration and size distribution. *Diam. Relat. Mater.* **2013**, *32*, 66–71. [CrossRef]
43. Mikheev, G.M.; Puzyr, A.P.; Vanyukov, V.V.; Mogileva, T.N.; Bondar, V.S. The effect of the size factor of nanodiamonds in suspensions on optical power limiting and nonlinear laser light scattering. *Tech. Phys. Lett.* **2013**, *39*, 229–232. [CrossRef]
44. Vanyukov, V.V.; Mikheev, G.M.; Mogileva, T.N.; Puzyr, A.P.; Bondar, V.S.; Svirko, Y.P. Concentration dependence of the optical limiting and nonlinear light scattering in aqueous suspensions of detonation nanodiamond clusters. *Opt. Mater.* **2014**, *37*, 218–222. [CrossRef]
45. Potamianos, D.; Papadakis, I.; Kakkava, E.; Bourlinos, A.B.; Trivizas, G.; Zboril, R.; Couris, S. Non-linear optical response of gold-decorated nanodiamond hybrids. *J. Phys. Chem. C* **2015**, *119*, 24614–24620. [CrossRef]
46. Muller, O.; Pichot, V.; Merlat, L.; Schmidlin, L.; Spitzer, D. Nonlinear optical behavior of porphyrin functionalized nanodiamonds: An efficient material for optical power limiting. *Appl. Opt.* **2016**, *55*, 3801–3808. [CrossRef]
47. Matshitse, R.; Khene, S.; Nyokong, T. Photophysical and nonlinear optical characteristics of pyridyl substituted phthalocyanine —Detonation nanodiamond conjugated systems in solution. *Diam. Relat. Mater.* **2019**, *94*, 218–232. [CrossRef]
48. Matshitse, R.; Nyokong, T. Substituent effect on the photophysical and nonlinear optical characteristics of Si phthalocyanine—Detonated nanodiamond conjugated systems in solution. *Inorganica Chim. Acta* **2020**, *504*, 119447. [CrossRef]
49. Mikheev, G.M.; Krivenkov, R.Y.; Mogileva, T.N.; Mikheev, K.G.; Nunn, N.; Shenderova, O.A. Saturable absorption in suspensions of single-digit detonation nanodiamonds. *J. Phys. Chem. C* **2017**, *121*, 8630–8635. [CrossRef]
50. Vanyukov, V.; Mikheev, G.; Mogileva, T.; Puzyr, A.; Bondar, V.; Lyashenko, D.; Chuvilin, A. Saturable absorption in detonation nanodiamond dispersions. *J. Nanophotonics* **2017**, *11*, 0325061–0325067. [CrossRef]
51. Krivenkov, R.Y.; Mikheev, K.G.; Mogileva, T.N.; Nunn, N.; Shenderova, O.A.; Mikheev, G.M. Optical limiting in nanodiamond suspension: Shortening of the laser pulses. *J. Phys. Conf. Ser.* **2018**, *1124*, 051017. [CrossRef]
52. Mikheev, G.M.; Mikheev, K.G.; Mogileva, T.N.; Puzyr, A.P.; Bondar, V.S. Laser image recording on detonation nanodiamond films. *Quantum Electron.* **2014**, *44*, 1–3. [CrossRef]
53. Street, T. Raman spectroscopy of amorphous, nanostructured, diamond-like carbon, and nanodiamond. *Phil. Trans. R. Soc. Lond. A* **2004**, *362*, 2477–2512.
54. Sun, Z.; Shi, J.R.; Tay, B.K.; Lau, S.P. UV Raman characteristics of nanocrystalline diamond films with different grain size. *Diam. Relat. Mater.* **2000**, *9*, 1979–1983. [CrossRef]
55. Nadolinny, V.A.; Yelisseyev, A.P. New paramagnetic centres containing nickel ions in diamond. *Diam. Relat. Mater.* **1993**, 3, 17–21. [CrossRef]
56. Aleksenskii, A.E.; Vul', A.Y.; Konyakhin, S.V.; Reich, K.V.; Sharonova, L.V.; Eidel'man, E.D. Optical properties of detonation nanodiamond hydrosols. *Phys. Solid State* **2012**, *54*, 578–585. [CrossRef]
57. Usoltseva, L.O.; Volkov, D.S.; Nedosekin, D.A.; Korobov, M.V.; Proskurnin, M.A.; Zharov, V.P. Absorption spectra of nanodiamond aqueous dispersions by optical absorption and optoacoustic spectroscopies. *Photoacoustics* **2018**, *12*, 55–66. [CrossRef]
58. Chapple, P.B.; Staromlynska, J.; Hermann, J.A.; Mckay, T.J.; Mcduff, R.G. Single-beam z-scan: Measurement techniques and analysis. *J. Nonlinear Opt. Phys. Mater.* **1997**, *6*, 251–293. [CrossRef]
59. Van Stryland, E.W.; Sheik-Bahae, M. Z-Scan technique for nonlinear materials characterization. *Proc. SPIE* **1997**, *10291*, 102910Q.
60. Mikheev, G.M.; Vanyukov, V.V.; Mogileva, T.N.; Puzyr', A.P.; Bondar', V.S.; Svirko, Y.P. Saturable absorption in aqueous suspensions of detonation nanodiamonds under irradiation with femtosecond laser pulses. *Tech. Phys. Lett.* **2015**, *41*, 1163–1166. [CrossRef]

61. Mansoor, S.-B.; Said, A.A.; Wei, T.-H.; Hagan, D.J.; van Stryland, E.W. Sensitive measurement of optical nonlinearities using a single beam. *IEEE J. Quantum Electron.* **1990**, *26*, 760–769.
62. Mikheev, G.M.; Kuznetsov, V.L.; Bulatov, D.L.; Mogileva, T.N.; Moseenkov, S.I.; Ishchenko, A.V. Photoinduced transparency of a suspension of onion-like carbon nanoparticles. *Tech. Phys. Lett.* **2009**, *35*, 162–165. [CrossRef]
63. Mikheev, G.M.; Kuznetsov, V.L.; Mikheev, K.G.; Mogileva, T.N.; Shuvaeva, M.A.; Moseenkov, S.I. Laser modification of optical properties of a carbon nanotube suspension in dimethylformamide. *Tech. Phys. Lett.* **2013**, *39*, 337–340. [CrossRef]
64. Bao, Q.; Zhang, H.; Wang, Y.; Ni, Z.; Yan, Y.; Shen, Z.X.; Loh, K.P.; Tang, D.Y. Atomic-layer craphene as a saturable absorber for ultrafast pulsed lasers. *Adv. Funct. Mater.* **2009**, *19*, 3077–3083. [CrossRef]
65. Yamashita, S.; Martinez, A.; Xu, B. Short pulse fiber lasers mode-locked by carbon nanotubes and graphene. *Opt. Fiber Technol.* **2014**, *20*, 702–713. [CrossRef]
66. Guo, B.; Xiao, Q.-L.; Wang, S.-H.; Zhang, H. 2D Layered materials: Synthesis, nonlinear optical properties, and device applications. *Laser Photonics Rev.* **2019**, *13*, 1800327. [CrossRef]
67. Svelto, O. *Principles of Lasers*; Tamburini: Milano, Italy, 1972.
68. Feng, Y.; Dong, N.; Wang, G.; Li, Y.; Zhang, S.; Wang, K.; Zhang, L.; Blau, W.J.J.; Wang, J. Saturable absorption behavior of free-standing graphene polymer composite films over broad wavelength and time ranges. *Opt. Express.* **2015**, *23*, 559–569. [CrossRef] [PubMed]
69. Koniakhin, S.V.; Rabchinskii, M.K.; Besedina, N.A.; Sharonova, L.V.; Shvidchenko, A.V.; Eidelman, E.D. Evidence of absorption dominating over scattering in light attenuation by nanodiamonds. *Phys. Rev. Res.* **2020**, *2*, 013316. [CrossRef]
70. Stehlik, S.; Varga, M.; Ledinsky, M.; Jirasek, V.; Artemenko, A.; Kozak, H.; Ondic, L.; Skakalova, V.; Argentero, G.; Pennycook, T.; et al. Size and purity control of HPHT nanodiamonds down to 1 nm. *J. Phys. Chem. C* **2015**, *119*, 27708–27720. [CrossRef]
71. Vul, A.Y.; Eydelman, E.D.; Sharonova, L.V.; Aleksenskiy, A.E.; Konyakhin, S.V. Absorption and scattering of light in nanodiamond hydrosols. *Diam. Relat. Mater.* **2011**, *20*, 279–284. [CrossRef]
72. Kryshtal, A.P.; Mchedlov-Petrossyan, N.O.; Laguta, A.N.; Kriklya, N.N.; Kruk, A.; Osawa, E. Primary detonation nanodiamond particles: Their core-shell structure and the behavior in organo-hydrosols. *Colloids Surf. A Physicochem. Eng. Asp.* **2021**, *614*, 126079. [CrossRef]
73. Yelisseyev, A.; Babich, Y.; Nadolinny, V.; Fisher, D.; Feigelson, B. Spectroscopic study of HPHT synthetic diamonds, as grown at 1500 C. *Diam. Relat. Mater.* **2002**, *11*, 22–37. [CrossRef]
74. Lindblom, J.; Holsa, J.; Papunen, H.; Hakkanen, H. Luminescence study of defects in synthetic as-grown and HPHT diamonds compared to nutural diamonds. *Am. Mineral.* **2005**, *90*, 428–440. [CrossRef]
75. Lu, H.-C.; Peng, Y.-C.; Lin, M.-Y.; Chou, S.-L.; Lo, J.-I.; Cheng, B.-M. Analysis of nickel defect in diamond with photoluminescence upon excitation near 200 nm. *Anal. Chem.* **2015**, *87*, 7340–7344. [CrossRef] [PubMed]

applied sciences

MDPI

Article

Carbon Dots with Up-Conversion Luminescence as pH Nanosensor

Kirill Laptinskiy [1,2], Maria Khmeleva [1], Alexey Vervald [1], Sergey Burikov [1,2] and Tatiana Dolenko [1,2,*]

1 Faculty of Physics, M.V. Lomonosov Moscow State University, Leninskie Gory 1/2, 119991 Moscow, Russia
2 D.V. Skobeltsyn Institute of Nuclear Physics, M.V. Lomonosov Moscow State University, Leninskie Gory 1/2, 119991 Moscow, Russia
* Correspondence: tdolenko@mail.ru

Abstract: In this study, the up-conversion luminescence for aqueous suspensions of carbon dots with polyfunctional and carboxylated surfaces synthesized by a hydrothermal method was investigated. The obtained quadratic dependence of the luminescence intensity on the power of the exciting radiation indicates that the up-conversion luminescence of these carbon dots is caused by two-photon absorption. The optimal wavelength of the exciting radiation was determined for the studied samples. The dependences of the signal for the up-conversion luminescence of carbon dots on the pH value of the suspension were obtained. It was shown that these carbon dots can be used as the nanosensor of pH of liquid media in a wide range of pH values. The advantage of this nanosensor is that the excitation of the up-conversion luminescence of carbon dots does not entail excitation of autoluminescence of the biological medium. It expands the possibilities of using this sensor in biomedical applications.

Keywords: carbon dots; up-conversion luminescence; two-photon absorption; nonlinear spectroscopy

Citation: Laptinskiy, K.; Khmeleva, M.; Vervald, A.; Burikov, S.; Dolenko, T. Carbon Dots with Up-Conversion Luminescence as pH Nanosensor. *Appl. Sci.* **2022**, *12*, 12006. https://doi.org/10.3390/app122312006

Academic Editor: Gennady M. Mikheev

Received: 31 October 2022
Accepted: 21 November 2022
Published: 24 November 2022

1. Introduction

In recent years, there has been a rapidly growing interest in the nonlinear optical properties of carbon nanoparticles [1–3]. This interest has been caused, among other things, by the wide prospects for the use of carbon dots (CDs) and nanodiamonds as medical nanoagent photoluminescent markers for the rapid and sensitive detection of analyte, drug carriers, adsorbents, etc. [4–8]. Together with the photostability and a high quantum yield of luminescence, solubility in water, biocompatibility, and the possibility of surface modification, the recently discovered property of various carbon dots to luminesce in the anti-Stokes region significantly expands the prospects for the use of such CDs in biomedicine. Upon excitation of the up-conversion luminescence (UCL) of nanoparticles, the autofluorescence of biological tissue is not excited, while UCL itself falls into the spectral region of the biological "transparency window", which ensures deep penetration of radiation in tissue.

The first report on the observation of UCL of carbon dots, upon the excitation of CDs obtained by laser ablation by pulsed laser radiation with a wavelength of 800 nm, was made by L. Cao et al. [9] in 2007. The CDs' UCL spectrum was a broad band with a maximum near 470 nm. The authors found a quadratic dependence of the UCL intensity of the CDs on the excitation power and explained the observed UCL as a result of two-photon absorption.

From 2007 to 2013, numerous publications appeared, including those in high-ranking journals, in which the authors reported on the observation of CDs' UCL when excited by a conventional xenon lamp of a spectrofluorimeter [10–12]. However, the authors of [13,14] showed that the supposed CD UCL signals observed on the spectrofluorimeter are artifacts such as interference from leaking second-order diffraction of the excitation light when broad-band light sources and diffraction gratings are used for excitation wavelength selection. According to [9,14,15], the main criterion for a "real" CD UCL is the quadratic dependence of the UCL intensity on the excitation radiation intensity.

The results of active studies on CDs' UCL over the past 5 years have shown that the properties of UCL of CDs of different synthesis differ significantly [9,14,16–18]. This is explained by the existence of various mechanisms for the formation of UCL [19], including UCL of CDs: two-photon absorption through real and virtual levels [9,14,15,20], multiphoton absorption [21], thermal activation [22], and the state of surface traps [20,23]. In the vast majority of publications, two-photon absorption is considered to be the main one.

However, the dependences of the CDs' UCL intensity on the power of the exciting radiation are not always quadratic. For example, for CDs doped with nitrogen by the photochemical method, these dependences resembled those for lanthanoid-based up-conversion nanoparticles and was linear in the double-log coordinates with the slope 0.69 [18], while for thermally-activated CDs, this dependence was found to be linear in standard linear coordinates (as well as in the double-log coordinates with the slope 1) [22].

The observed dependences of the CDs' UCL intensity on the excitation wavelength also differ. For CDs synthesized by different methods or doped with different atoms, these dependences have one [21], two [18,24], or three [25] maxima with different positions. For other CDs, such as CDs simultaneously doped with S and Se [16] and for CDs synthesized by the Hummers method [17], excitation wavelength-independent NIR emission was observed.

Thus, the accumulated results of UCL studies of various CDs at this stage are characterized by many contradictions and uncertainty. The next step is to search for correlations and systematize all the data obtained. It should also be noted that CDs' UCL with different surface functionalization has not yet been studied, although it is well known that the surface properties of nanoparticles significantly affect their properties, primarily colloidal and optical [26,27].

Despite the contradictions in the results of studying the properties of CDs' UCL, there are many examples of successful application of CDs with UCL in various fields. The main application of CDs with UCL is the optical visualization of nanoparticles in cancer cells and biological tissues [9,28], in a multiphoton confocal microscope [18], for photothermal therapy of cancerous tumors [12]. It was shown that CDs' UCL could be detected in tissue from a depth up to 1800 μm [21], which is an important result from the point of view of CDs applicability in biological imaging. The authors of [22] suggested using CDs with UCL for thermosensing. Due to the fact that some substances quench UCL upon interaction with CDs, it was proposed to use CDs with UCL as nanosensors for hypochlorite [29], hydrogen sulfide [30], and Cu^{2+} ions [31]. CDs with UCL can be used as a laser amplification medium in low-threshold diode-pumped lasers [32].

While pH nanosensors on the basis of CDs' Stokes luminescence have been actively developed [8,33,34], there are almost no publications devoted to the CDs' UCL dependence on pH. At the time of writing, we managed to find only one paper, in which such dependence was shown: intensity of UCL (emission with the maximum around 640 nm upon the excitation at 690 nm with pulse laser) for the studied CDs was shown to significantly change in cells upon their incubation at buffers with different pH (from 4 to 8), at tumor/normal muscle mice tissue (pH 6.5–6.8 against pH 7.4), and in zebrafish injected with CDs in buffers of different pH (from 4 to 8). Unfortunately, while this publication presents a dependence on the Stokes luminescence spectra for CDs on pH in wide region 0.5–9 (excitation at 573 nm), the same was not the case for their UCL [35]. The advantage of this nanosensor, based on UCL is that excitation of up-conversion luminescence does not lead to autoluminescence of the biological medium.

This article presents the results of studying the dependence of the intensity of UCL on the wavelength and on the power of the exciting radiation for different functionalized CDs synthesized by the hydrothermal method. The features of CDs' UCL with a functionalized surface were studied for the first time. The possibility of using CDs with UCL as a pH nanosensor is shown. For the first time, the accuracy of determining the pH of an aqueous solution and biological tissue using the CDs' UCL spectra was estimated.

2. Materials, Their Characterization, and Research Methods

2.1. Synthesis of Carbon Dots and Functionalization of Their Surface

The objects of study were CDs obtained by the hydrothermal method from citric acid and ethylenediamine. The concentrations of precursors in the initial aqueous solution were 0.15 M for citric acid and 3 M for ethylenediamine (i.e., the ratio of precursors was 1:20, respectively). This solution was placed in a 15 mL polytetrafluoroethylene (PTFE) tube, which was placed in an autoclave. The autoclave with a mixture of precursors was placed in a Sputnik muffle furnace (Russia). The heating temperature was 190 °C, the heat treatment time was 2 h. Next, the autoclave was removed from the oven and cooled at room temperature for 7 h. To remove large particles, the synthesized CDs were passed through a syringe membrane filter with a pore diameter of 0.22 μm.

In this way, CDs with a polyfunctional surface (CD-poly) were synthesized, which were then functionalized with hydroxyl (CD-OH) and carboxyl (CD-COOH) groups. To functionalize CDs with hydroxyl groups (CD-OH), the dried residue of the solution after synthesis was mixed with 1 g of sodium nitride and 3 mL of hydrochloric acid to remove amide groups. To obtain the carboxylated surface of CDs (CD-COOH), 10 mL of water and 4.2 mL of NaOH solution with a concentration of 0.5 M were added to the dried residue.

2.2. Electron Microscopy of Obtained Carbon Dots

Studies of the CDs' structure were carried out by scanning transmission electron microscopy using a high-angle annular dark field detector (HAADF STEM) on a Tecnai Osiris microscope (Thermo Fisher Scientific, USA) at an accelerating voltage of 200 kV. The results are shown in Figure 1a. Elemental analysis was carried out using a special SuperX EDS system, which includes four Bruker silicon detectors, the design of which makes it possible to obtain the distribution of chemical elements in a few minutes. The diagram of elemental analysis is presented in Figure 1b.

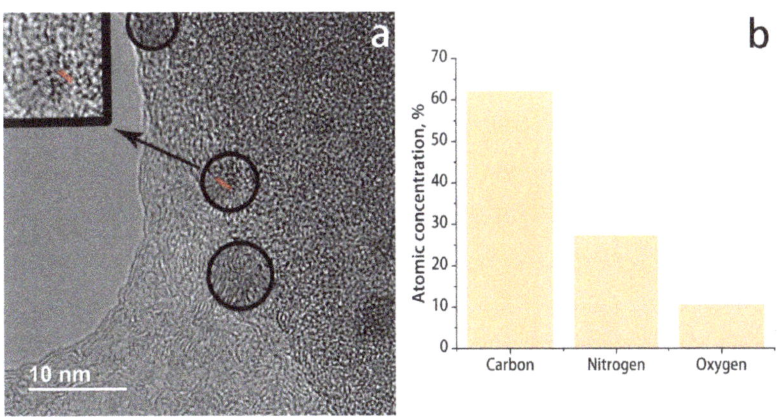

Figure 1. STEM image of the synthesized nanoparticles (**a**) and their elemental analysis (**b**).

As can be seen from the presented image, the synthesized nanoparticles are about 7 nm in size. The lattice fringes are clearly visible (Figure 1, inset, red lines) with the lattice parameter around 0.189 nm that corresponds to the (102) lattice fringes of graphite.

2.3. Preparation and Characterization of Aqueous Solutions of Carbon Dots

Deionized water (Millipore Simplicity UV water purification system) was used to prepare aqueous solutions of CDs. Aqueous solutions of CD-poly, CD-COOH, and CD-OH were prepared with a nanoparticle concentration of 0.1 mg/mL (obtained from gravimetric analysis).

To change the pH in the range from 2 to 12, either an aqueous solution of HCl (Sigma Aldrich, concentration 1 M, pH = 0) or an aqueous solution of NaOH (Dia-M, concentration 1.8 M, pH = 14) was added to the solution. Measurements for the pH value of aqueous solutions of CDs were carried out using the ionometric converter Akvilon I-500 equipped with an InLab Nano pH electrode (Mettler Toledo). All measurements were carried out at a fixed temperature of 22 °C.

Zeta potentials and sizes of nanoparticles in aqueous solutions were measured using dynamic light scattering on a Malvern Zetasizer NanoZS instrument. According to the data obtained, the sizes of the studied CDs were 182 ± 21, 105 ± 12, and 176 ± 23 nm for CD-poly, CD-COOH, and CD-OH, respectively. The corresponding zeta potentials were −39.7 ± 4.3, −29.1 ± 4.5, and −17.2 ± 1.3 mV. The obtained values of the zeta potentials indicate the colloidal stability of aqueous solutions of all CDs.

2.4. IR Absorption Spectroscopy of Carbon Dots

Analysis of the structure and functional surface groups of CDs was carried out using FTIR spectroscopy on a Bruker Invenio R IR spectrometer equipped with an ATR attachment based on a diamond crystal. The spectral resolution was 4 cm^{-1}. The obtained IR absorption spectra for the studied CDs are shown in Figure 2.

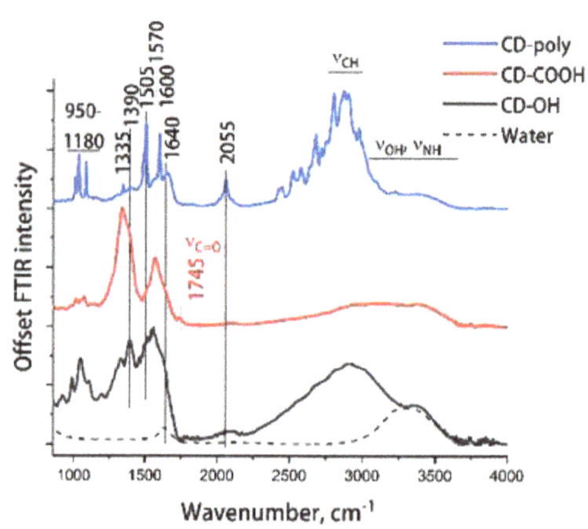

Figure 2. IR absorption spectra of water and powders of synthesized CDs.

The IR absorption spectra of three CD samples contain a similar set of bands that differ in relative absorption intensity. In the region of 3000–3700 cm^{-1}, there is a wide band caused by stretching vibrations of the surface hydroxyl and amino groups and OH groups of water molecules adsorbed on the surface of the samples. The relative intensity of this band for CD-poly and CD-COOH is significantly lower than for CD-OH. In the region of 2800–3000 cm^{-1} there are bands of stretching vibrations of CH_x groups. In the region of 2055 cm^{-1}, there is a band that presumably belongs to overtones/combinations of various vibrations in CDs [36].

The band at 1745 cm^{-1} in the spectrum of CD-COOH corresponds to the stretching vibrations of C=O bonds of surface carboxyl groups. The weak relative intensity of this band is due to the fact that the COOH groups belong exclusively to the surface of this CD, while most other vibrational bands are due to the vibrations of the groups that are also in the CD's core. This situation is typical for CDs with a size of about 100 nm and does not manifest itself, for example, for nanodiamonds with a carboxylated surface: in them, the diamond core is not active, and the relative intensity of the vibration bands of all surface groups is much higher [37].

In the region of 1640 cm^{-1}, there is a band of bending vibrations of OH bonds. In the spectra of CD-COOH and CD-OH, it manifests itself as a "shoulder" of a wide intense band with a maximum in the region of 1570 cm^{-1}, which is due to stretching vibrations of C=C bonds [36]. In the region of 1580–1650 cm^{-1}, the band of bending vibrations of NH groups overlaps with it.

In the IR spectrum of the CD-poly, intense narrow bands are observed with maxima in the regions of 1600 and 1505 cm^{-1}, which, according to the authors of [36,38], are due to skeletal vibrations of the polycyclic aromatic elements of CDs. The absence of such bands in the spectra of CD-COOH and CD-OH indicates that the functionalization of these CDs occurred primarily due to the rupture and subsequent addition of radicals to such aromatic rings.

The band with a maximum in the region of 1390 cm^{-1} corresponds to stretching vibrations of C–O bonds in –COOH groups [39]. The band with a maximum in the region of 1335 cm^{-1} is also due to C–O vibrations, but particularly in the C–O–C groups [40]. The set of bands in the range of 950–1180 cm^{-1} is mainly due to stretching vibrations of C–O and bending vibrations of C–H bonds in various environments [36].

An analysis of the IR spectra showed that the CD-poly contains a large number of polycyclic aromatic elements and various functional groups, such as C–O–C, C=C, N–H, OH. As a result of the functionalization of the CD-poly, on the surfaces of CD-COOH and CD-OH during the destruction of aromatic rings, a large number of oxygen-containing groups were formed, and the relative intensity of the bands due to vibrations of the C–O–C, C–O, and OH groups bonds increased. Additionally, in the spectra of CD-COOH bands corresponding to vibrations of carboxyl groups appeared. Thus, the results of IR absorption spectroscopy allow us to conclude that the functionalization of CD-COOH and CD-OH was successful (Figure 2).

2.5. Spectrophotometry of Aqueous Solutions of CDs

The optical extinction spectra of aqueous solutions of CD-poly, CD-COOH, and CD-OH were obtained on a Shimadzu UV-1800 spectrophotometer in the wavelength range of 190–1100 nm with a resolution of 1 nm. The informative part of the obtained spectra is shown in Figure 3.

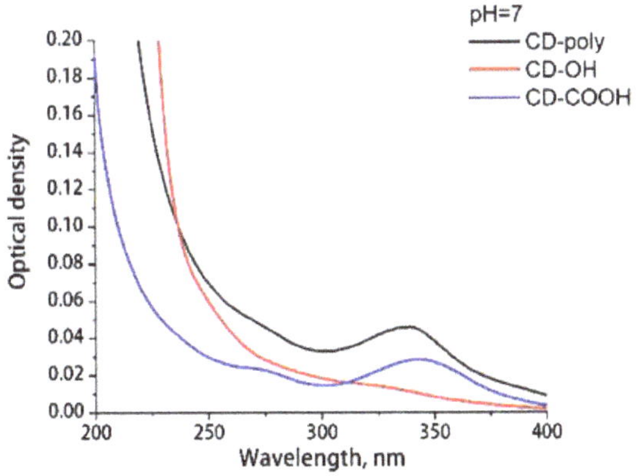

Figure 3. Optical extinction spectra of aqueous solutions for the studied CDs. pH = 7.

As can be seen from the presented data, the extinction spectra of aqueous solutions of CDs have a number of features. In all optical extinction spectra for the studied aqueous solutions, an intense peak is observed in the short-wavelength region, the right shoulder of the band of which extends up to ~250–300 nm. According to the literature data [41], the peak of CDs in this region is due to π–π* transitions in aromatic sp^2 carbons. In the extinction spectra of CDs with polyfunctional and carboxylated surfaces, a peak is observed in the region of 275 nm, corresponding to n–π* transitions of C=O [42] bonds and corresponding to n–π* transitions of C=O, C–N, or C–OH bonds in sp^3-hybridized domains associated with carboxyl (–COOH) or amine ($–NH_2$) groups on the CDs' surface [43]. In the optical extinction spectra of CD-OH, these peaks are absent.

2.6. Photoluminescence Spectroscopy

The up-conversion luminescence of the CDs was studied on the experimental setup shown in Figure 4. A pulsed nanosecond Nd:YAG laser (model LQ629-100, Solar Laser Systems, Belarus) was used as a pump source. The third harmonic (355 nm, pulse duration 10 ns, pulse repetition rate 100 Hz) of the Nd:YAG laser was used to pump an LP603 parametric light generator (Solar Laser Systems, Belarus), whose radiation was the source of excitation of the CDs' UCL. This laser system allows the pump wavelength to be changed in the range from 532 to 1070 nm. The radiation of the parametric light generator passed through a KS-11 light filter, which did not transmit the third harmonic of the initial laser radiation, and was directed through a rotating mirror to a mechanical optical shutter, which was used to record and then subtract the background signal from the CDs' UCL spectra. The probing of the sample in the cuvette was carried out through the bottom of the cuvette. The resulting radiation of the sample at an angle of 90° was collected with a lens system to a recording system consisting of a monochromator (MBP80) with a focal length of 500 mm, equipped with a grating of 150 lines/mm, and a CCD camera (Horiba-Jobin Yvon, Synapse BIUV). Light filters SZS-22 and SZS-23 at the entrance of monochromator did not transmit the signal from the pump and isolated the luminescent signal in the spectral range of registration from 418 to 773 nm. The spectral resolution of the experimental setup was 2 nm. The power of the excitation radiation at the sample was recorded using an Ophir Nova 2 power meter equipped with a PE50-DIF-C pyroelectric sensor.

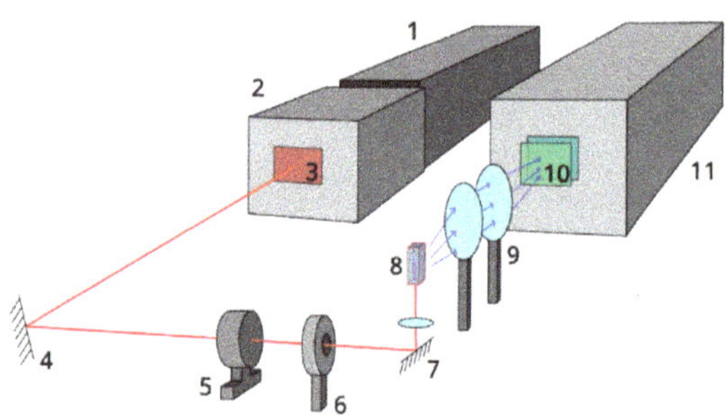

Figure 4. Scheme of the experimental setup for the study of CDs' UCL: 1—pulsed nanosecond Nd:YAG laser; 2—parametric light generator; 3—light filter KS-11; 4—rotatory mirror; 5—power meter; 6—mechanical shutter; 7—rotatory mirror and lens; 8—cuvette with the solution of nanoparticles; 9—optical system of two lenses; 10—light filters SZS-22 and SZS-23; 11—registration system consisting of a monochromator and a CCD camera.

3. Results and Discussion

3.1. Dependence of the UCL Intensities of Aqueous Solutions of CDs on the Wavelength of the Exciting Radiation

To study the UCL of CDs, the UCL spectra of CDs' aqueous solutions at pH 11.7 (initial pH for studied aqueous solutions) were obtained under excitation by laser radiation with different wavelengths in the range from 670 to 840 nm with excitation radiation power of 50 mW (Figure 5). From them, the dependences of the integral intensities of these spectra on the wavelength of the exciting radiation were calculated (Figure 6). For CD-OH, no photoluminescence signal was observed in the specified spectral range, what corresponds to the absence of the intense absorption bands in the optical absorption spectra of this sample in the range from 250 to 500 nm. Additionally, it is worth noting that the synthesized CDs were characterized by stable UCL, both in terms of reproducibility of the up-conversion luminescence of CD aqueous suspension obtained in different synthesis batches under the same synthesis conditions (see Figure S1) and in terms of the UCL stability of aqueous suspensions of CDs for 10 days (see Figure S2).

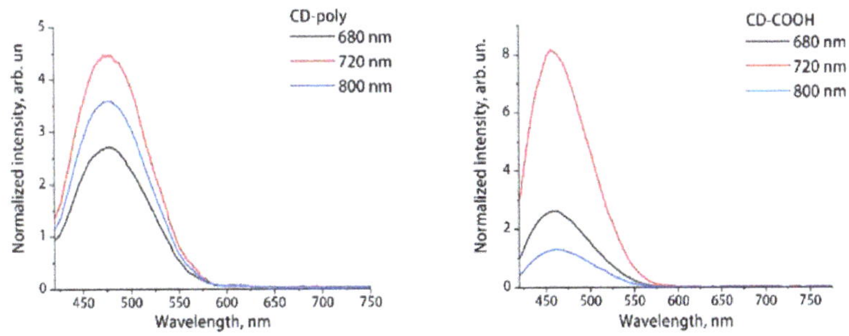

Figure 5. UCL spectra of aqueous solutions of CD-poly and CD-COOH upon excitation by laser radiation at wavelengths of 680, 720, and 800 nm (pH = 11.7).

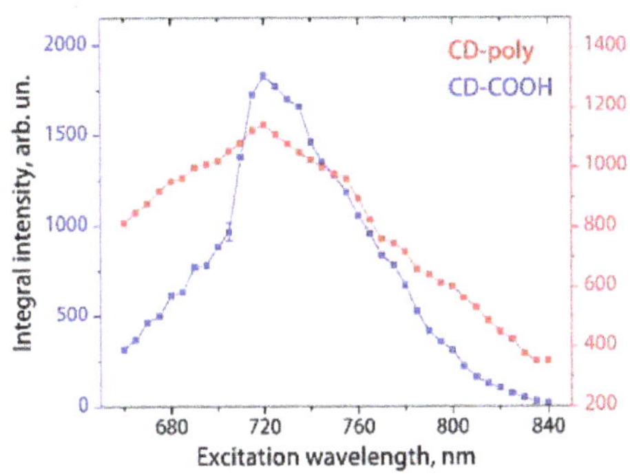

Figure 6. Dependences of the integrated UCL intensities of aqueous solutions of CD-poly and CD-COOH on the wavelength of the exciting radiation (pH = 11.7).

The processing of the spectra consisted in averaging 4 consecutively recorded spectra and normalizing their intensity to the power of the exciting radiation and to the time of accumulation of the spectra. The spectra were smoothed by a third-order polynomial using the Savitsky–Golay method.

It can be seen from the obtained results that the maxima of the dependences of the UCL integral intensity on the wavelength of the exciting radiation for the two types of CDs are close and are in the region of 720 nm (Figure 6). Therefore, this excitation wavelength can be considered optimal for both types of CDs, and it was used for the excitation of the CDs' UCL in following measurements. The proximity of the excitation wavelengths that provide the maximum integrated intensity of UCL of CD-COOH and CD-OH is explained by the same adsorption features of these CDs' solutions related to their functional groups (Figure 3). Nevertheless, it should be noted that, upon excitation of UCL by radiation at a wavelength of 720 nm, the UCL intensity of an aqueous solution of CD-COOH is approximately 1.7 times higher than that of an aqueous solution of CD-poly at pH = 11.7. Thus, the presence/absence of CDs' UCL and its characteristics depend on the functionalization of the nanoparticle surface.

3.2. Dependence of the UCL Intensity of Aqueous Solutions of CDs on the Power of the Exciting Radiation

As discussed earlier, according to some photophysical models, the UCL of CDs is due to a two-photon absorption process. In this case, the CDs' UCL integrated intensity should depend quadratically on the power of the exciting radiation.

To check this, the UCL spectra of aqueous solutions of CD-poly and CD-COOH were obtained for different values of the excitation radiation power, which varied in the range from 12 mW ($5.1 \cdot 10^4$ J$\cdot$cm$^{-2}\cdot$s^{-1}) to 83 mW ($3.5 \cdot 10^5$ J$\cdot$cm$^{-2}\cdot$s^{-1}). The dependences of the UCL integrated intensity of CD-poly and CD-COOH in water on the value of the pump power are shown in Figure 7. They are well approximated by a quadratic function.

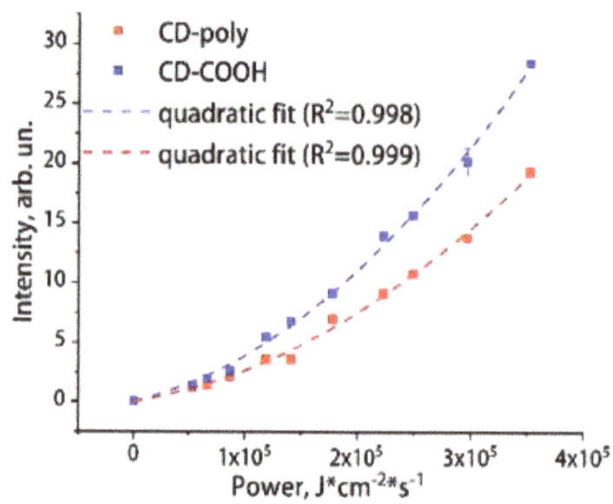

Figure 7. Dependences of the UCL integrated intensity of aqueous solutions of CD-poly and CD-COOH (pH = 11.7) on the power of exciting radiation and approximation of experimental data by quadratic dependences.

The obtained quadratic dependences for the intensity of UCL in aqueous solutions of CDs on the power of the exciting radiation prove the fact that the UCL of studied CDs is due to two-photon absorption. In following processing of the spectral data, the intensity of the CDs' UCL will be normalized to the square of the power of the exciting radiation.

From Sections 3.1 and 3.2 it can be concluded that the mechanisms for nonlinear (photoluminescence in the result of two-photon absorption) optical properties of CDs, as with linear ones (classic photoluminescence in the Stokes region [34,44]), depend on the type and composition of surface functional groups.

3.3. Dependence of the UCL Intensity of Aqueous Solutions of CDs on pH

The UCL spectra for aqueous solutions of CD-poly and CD-COOH with pH from 2 to 12 were obtained. From them, the dependences of the integrated intensity and position of the maxima of UCL spectra of CDs on the pH of aqueous solutions were calculated (Figure 8). As can be seen, these spectral parameters depend significantly on pH. Moreover, for CDs with polyfunctional and carboxylated surfaces, these dependences differ significantly. The UCL intensity of an aqueous solution of CD-COOH monotonically increases as the pH of the solution increases from 2 to 12, while the UCL intensity of CD-poly increases with the increase in pH from 2 to 5, and then decreases at pH > 5 (Figure 8a).

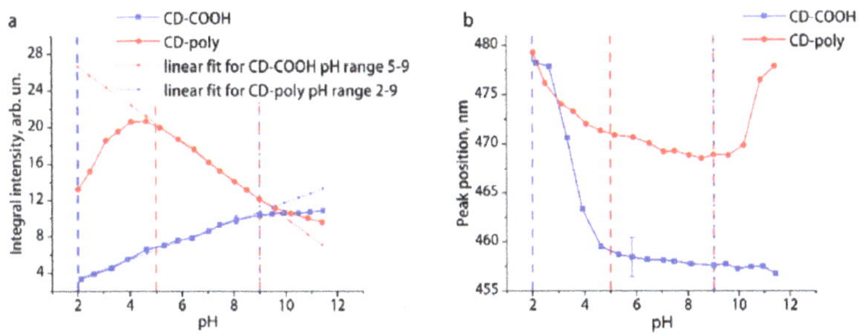

Figure 8. Dependence of the integral intensity (**a**) and the position of the maximum (**b**) of the UCL spectra of aqueous solutions of CD-COOH and CD-poly on pH. The vertical lines represent the pH limits of applying the linear fit to UCL intensity vs. pH.

The maximum of the UCL spectra of CD-COOH sharply shifts to the short-wavelength region by 20 nm with an increase in the pH value of the solution from 2 to 5, while at further increase in the pH its position practically does not change (Figure 8b). It should be noted that in the pH range from 2 to 5, deprotonation of carboxyl groups on the surface of nanoparticles in water occurs, which significantly affects the state of the electronic levels of surface states [37,44]. The maximum UCL spectra of CD-poly gradually shifts to the short wavelength region by 10 nm with an increase in the pH value of the solution from 2 to 10, and then sharply shifts by 8 nm to the long wavelength region with a further increase in the pH of the solution. It can be assumed that the course of the dependences obtained for CD-poly solutions is also to some extent affected by the processes of deprotonation in the COOH, OH, and NH surface groups in water [37,44].

3.4. Carbon Dots with Up-Conversion Luminescence as a pH Nanosensor

It follows from the obtained results that the dependences of the integrated intensity for UCL of CDs' aqueous solutions on the pH have linear sections: CD-COOH in the pH range from 2 to 9, CD-poly in the pH range from 5 to 9 (Figure 8a). This means that the studied CDs with UCL can serve as promising nanoagents for use as nanosensors of pH in liquid media, including at the cellular level.

Approximations of the linear regions for the dependences of the UCL integrated intensity of aqueous solutions of CDs on pH (Figure 8a) can be used as calibration curves for the determination of the pH of the CDs' aqueous environment. Using them, the pH of the environment can be measured from the UCL spectra of CD-poly with an accuracy of 0.13, and from the UCL spectra of CD-COOH with an accuracy of 0.21.

Approbation of the pH nanosensor based on the CDs' UCL was carried out for chicken egg white (Figure 9). For this, the UCL spectra of CDs added to four different egg whites (concentration of CDs was 0.1 mg/mL, the same as for aqueous solutions) were obtained, and the integrated intensity of the UCL of all spectra was determined. For each egg white, five UCL spectra of CD-poly and CD-COOH were obtained. Using the calibration curves of the linear sections of the dependences of the integral intensity of UCL on pH (Figure 8a), the pH values for each sample were determined, and averaged for five corresponding spectra. Parallel to it, the pH value at the probing point of the exciting laser radiation of each chicken egg white was measured using a pH electrode. All of the obtained pH values are presented in Table 1.

Figure 9. UCL of CDs in chicken egg white (lower half of the image) upon excitation at a wavelength of 720 nm.

Table 1. The pH values of the water and chicken egg white obtained using a pH electrode and from the spectra of CDs' UCL.

Measurement Method	Water	Egg White 1	Egg White 2	Egg White 3	Egg White 4
Electrode	6.73 ± 0.05	8.33 ± 0.05	8.36 ± 0.05	8.23 ± 0.05	8.44 ± 0.05
CD-poly	6.74 ± 0.12	8.34 ± 0.16	8.32 ± 0.18	8.16 ± 0.19	8.48 ± 0.16
CD-COOH	6.73 ± 0.22	8.35 ± 0.31	8.33 ± 0.33	8.17 ± 0.368	8.55 ± 0.33

As can be seen from the presented results, the pH values of egg whites obtained by two independent methods are in good agreement. The error of the determination of pH for the CD-poly samples is smaller than for the CD-COOH samples, what was already observed for CDs' aqueous suspensions. In addition, the slope of the calibrations' straight line for CD-poly is greater in modulus than for CD-COOH (Figure 6), which also affects the accuracy of determining the pH of the medium. Thus, the functionalization of the CDs' surface with carboxyl groups did not give a benefit in pH measurement. However, it is known that it is the carboxylation of the surface of carbon nanoparticles that provides them with a high degree of biocompatibility, which is extremely important when using CDs in nanomedicine.

It should be noted that the accuracy of pH measurement according to the UCL spectra of CDs in aqueous solution is higher than that in chicken egg white. This is explained by the fact that the calibration dependences of the UCL intensity on pH obtained for an aqueous solution of CDs were used to determine the pH in the egg white. Nevertheless, the accuracy of determining the pH of both water and biological tissue from the UCL spectra of the nanosensor based on CDs (Table 1) satisfies the needs of biomedicine.

Thus, the pH nanosensor on the basis on CDs' up-conversion luminescence intensity was successfully approbated for the biological substances. Due to the fact that the anti-Stokes luminescence does not overlap with the autofluorescence of the biological medium, a good accuracy of pH determination was achieved. All pH nanosensors based on Stokes luminescence would be hindered by the biological autofluorescence, fully avoided for UCL-based probes.

4. Conclusions

In this work, the UCL for aqueous solutions of carbon dots with a polyfunctional surface and a surface functionalized with carboxyl groups was studied. It was found that UCL is absent in CDs functionalized with hydroxyl surface groups.

The obtained quadratic dependence of the CDs' UCL intensity on the power of the exciting radiation proves the fact that the UCL for the studied CDs is caused by two-photon absorption processes.

The differences in the UCL of CD-poly and CD-COOH aqueous solutions and its absence for CD-OH indicates that the mechanisms of nonlinear optical properties of CDs (photoluminescence in the result of two-photon absorption), such as linear ones (classic photoluminescence in the Stokes region), depend on the type and composition of surface functional groups.

It was found that the UCL for aqueous solutions of CD-poly and CD-COOH in water depends on the wavelength of the excitation radiation and on the pH value of the solution. The analysis of the obtained results showed that CDs' UCL is most effectively excited at a wavelength of 720 nm. It has been shown that the studied CD-COOH and CD-poly with up-conversion luminescence can be used as pH nanosensors in the linear response pH ranges from 2 to 9 and from 5 to 9, respectively. In these ranges, the estimated accuracy in the determination of the pH of the environment by UCL spectra for aqueous solutions of CDs was 0.13 for CD-poly and 0.21 for CD-COOH. Approbation of pH nanosensors in biological tissue—chicken egg white—was carried out, which demonstrated the prospects of using the proposed method of using CDs' UCL to determine the pH of biological media.

Obviously, CD-COOH are more promising for biomedical problems as pH nanosensors, since they have high biocompatibility and can measure pH in a larger range of its linear changes in biological tissue.

Supplementary Materials: The following supporting information can be downloaded at: https://www.mdpi.com/article/10.3390/app122312006/s1, Figure S1: Reproducibility of up-conversion luminescence of CD aqueous suspension obtained in different synthesis batches under the same synthesis conditions, Figure S2: Demonstration of the up-conversion luminescence stability of aqueous suspension of CDs for 10 days.

Author Contributions: Conceptualization, T.D. and K.L.; methodology, S.B., M.K., and K.L.; software, A.V.; validation, M.K. and A.V.; formal analysis, A.V.; investigation, M.K. and K.L.; resources, T.D.; data curation, T.D. and K.L.; writing—original draft preparation, T.D. and K.L.; writing—review and editing, T.D.; visualization, M.K.; supervision, T.D.; project administration, T.D.; funding acquisition, T.D. All authors have read and agreed to the published version of the manuscript.

Funding: The study was carried out at the expense of a grant from the Russian Science Foundation No. 22-12-00138, https://rscf.ru/project/22-12-00138/ (accessed on 2 September 2022).

Institutional Review Board Statement: Not applicable.

Informed Consent Statement: Not applicable.

Data Availability Statement: Not applicable.

Acknowledgments: We acknowledge the Interdisciplinary Scientific and Educational School of Lomonosov Moscow State University "Photonic and Quantum technologies. Digital medicine". Some of the experimental results used in this study were obtained using a FTIR spectrometer purchased under the Development Program of Moscow State University (Agreement No. 65, 4 October 2021).

Conflicts of Interest: The authors declare no conflict of interest.

References

1. Mansuriya, B.D.; Altintas, Z. Carbon Dots: Classification, Properties, Synthesis, Characterization, and Applications in Health Care—An Updated Review (2018–2021). *Nanomaterials* **2021**, *11*, 2525. [CrossRef]
2. Mikheev, G.M.; Vanyukov, V.V.; Mogileva, T.N.; Mikheev, K.G.; Aleksandrovich, A.N.; Nunn, N.A.; Shenderova, O.A. Femtosecond Optical Nonlinearity of Nanodiamond Suspensions. *Appl. Sci.* **2021**, *11*, 5455. [CrossRef]
3. Mikheev, G.M.; Krivenkov, R.Y.; Mogileva, T.N.; Mikheev, K.G.; Nunn, N.; Shenderova, O.A. Saturable absorption in suspensions of single-digit detonation nanodiamonds. *J. Phys. Chem. C* **2017**, *121*, 8630–8635. [CrossRef]
4. Sarmanova, O.E.; Burikov, S.A.; Dolenko, S.A.; Isaev, I.V.; Laptinskiy, K.A.; Prabhakar, N.; Karaman, D.Ş.; Rosenholm, J.M.; Shenderova, O.A.; Dolenko, T.A. A Method for Optical Imaging and Monitoring of the Excretion of Fluorescent Nanocomposites from the Body Using Artificial Neural Networks. *Nanomed. Nanotechnol. Biol. Med.* **2018**, *14*, 1371–1380. [CrossRef] [PubMed]
5. Zhang, H.; Wu, S.; Xing, Z.; Wang, H.-B. Turning waste into treasure: Chicken eggshell membrane derived fluorescent carbon nanodots for the rapid and sensitive detection of Hg2+ and glutathione. *Analyst* **2021**, *146*, 7250–7256. [CrossRef]
6. Zhang, H.; Wu, S.; Xing, Z.; Gao, M.; Sun, M.; Wang, J.; Wang, H.-B. Green synthesis of carbon nanodots for direct and rapid determination of theophylline through fluorescence turn on–off strategy. *Appl. Phys. A* **2022**, *128*, 1–10. [CrossRef]

7. Laptinskiy, K.; Burikov, S.; Dolenko, S.; Efitorov, A.; Sarmanova, O.; Shenderova, O.; Vlasov, I.; Dolenko, T. Monitoring of Nanodiamonds in Human Urine Using Artificial Neural Networks. *Phys. Status Solidi A* **2016**, *213*, 2614–2622. [CrossRef]
8. Sarmanova, O.E.; Laptinskiy, K.A.; Khmeleva, M.Y.; Burikov, S.A.; Dolenko, S.A.; Tomskaya, A.E.; Dolenko, T.A. Development of the Fluorescent Carbon Nanosensor for PH and Temperature of Liquid Media with Artificial Neural Networks. *Spectrochim. Acta A Mol. Biomol. Spectrosc.* **2021**, *258*, 119861. [CrossRef]
9. Cao, L.; Wang, X.; Meziani, M.J.; Lu, F.; Wang, H.; Luo, P.G.; Lin, Y.; Harruff, B.A.; Veca, L.M.; Murray, D.; et al. Carbon Dots for Multiphoton Bioimaging. *J. Am. Chem. Soc.* **2007**, *129*, 11318–11319. [CrossRef]
10. Zhu, S.; Zhang, J.; Tang, S.; Qiao, C.; Wang, L.; Wang, H.; Liu, X.; Li, B.; Li, Y.; Yu, W.; et al. Surface Chemistry Routes to Modulate the Photoluminescence of Graphene Quantum Dots: From Fluorescence Mechanism to Up-Conversion Bioimaging Applications. *Adv. Funct. Mater.* **2012**, *22*, 4732–4740. [CrossRef]
11. Shen, J.; Zhu, Y.; Chen, C.; Yang, X.; Li, C. Facile Preparation and Upconversion Luminescence of Graphene Quantum Dots. *Chem. Commun.* **2011**, *47*, 2580–2582. [CrossRef] [PubMed]
12. Zhuo, S.; Shao, M.; Lee, S.-T. Upconversion and Downconversion Fluorescent Graphene Quantum Dots: Ultrasonic Preparation and Photocatalysis. *ACS Nano* **2012**, *6*, 1059–1064. [CrossRef]
13. Tan, D.; Zhou, S.; Qiu, J. Comment on "Upconversion and Downconversion Fluorescent Graphene Quantum Dots: Ultrasonic Preparation and Photocatalysis". *ACS Nano* **2012**, *6*, 6530–6531. [CrossRef] [PubMed]
14. Gan, Z.; Wu, X.; Zhou, G.; Shen, J.; Chu, P.K. Is There Real Upconversion Photoluminescence from Graphene Quantum Dots? *Adv. Opt. Mater.* **2013**, *1*, 554–558. [CrossRef]
15. Wen, X.; Yu, P.; Toh, Y.-R.; Ma, X.; Tang, J. On the Upconversion Fluorescence in Carbon Nanodots and Graphene Quantum Dots. *Chem. Commun.* **2014**, *50*, 4703–4706. [CrossRef] [PubMed]
16. Lan, M.; Zhao, S.; Zhang, Z.; Yan, L.; Guo, L.; Niu, G.; Zhang, J.; Zhao, J.; Zhang, H.; Wang, P.; et al. Two-Photon-Excited near-Infrared Emissive Carbon Dots as Multifunctional Agents for Fluorescence Imaging and Photothermal Therapy. *Nano Res.* **2017**, *10*, 3113–3123. [CrossRef]
17. Santos, C.I.M.; Mariz, I.F.A.; Pinto, S.N.; Gonçalves, G.; Bdikin, I.; Marques, P.A.A.P.; Neves, M.G.P.M.S.; Martinho, J.M.G.; Maçôas, E.M.S. Selective Two-Photon Absorption in Carbon Dots: A Piece of the Photoluminescence Emission Puzzle. *Nanoscale* **2018**, *10*, 12505–12514. [CrossRef]
18. Jin, Q.; Gubu, A.; Chen, X.; Tang, X. A Photochemical Avenue to Photoluminescent N-Dots and Their Upconversion Cell Imaging. *Sci. Rep.* **2017**, *7*, 1793. [CrossRef] [PubMed]
19. Joly, A.G.; Chen, W.; McCready, D.E.; Malm, J.-O.; Bovin, J.-O. Upconversion Luminescence of CdTe Nanoparticles. *Phys. Rev. B* **2005**, *71*, 165304. [CrossRef]
20. Gui, R.; Jin, H.; Wang, Z.; Tan, L. Recent Advances in Optical Properties and Applications of Colloidal Quantum Dots under Two-Photon Excitation. *Coord. Chem. Rev.* **2017**, *338*, 141–185. [CrossRef]
21. Liu, K.; Song, S.; Sui, L.; Wu, S.; Jing, P.; Wang, R.; Li, Q.; Wu, G.; Zhang, Z.; Yuan, K.; et al. Efficient Red/Near-Infrared-Emissive Carbon Nanodots with Multiphoton Excited Upconversion Fluorescence. *Adv. Sci.* **2019**, *6*, 1900766. [CrossRef] [PubMed]
22. Li, D.; Liang, C.; Ushakova, E.V.; Sun, M.; Huang, X.; Zhang, X.; Jing, P.; Yoo, S.J.; Kim, J.; Liu, E.; et al. Thermally Activated Upconversion Near-Infrared Photoluminescence from Carbon Dots Synthesized via Microwave Assisted Exfoliation. *Small* **2019**, *15*, 1905050. [CrossRef] [PubMed]
23. Kumar Reddy Bogireddy, N.; Agarwal, V. Tunable Upconversion Emission from Oil-Based Carbon Nanodots. *Mater. Lett.* **2022**, *313*, 131640. [CrossRef]
24. Zhang, Q.; Wang, R.; Feng, B.; Zhong, X.; Ostrikov, K. Photoluminescence Mechanism of Carbon Dots: Triggering High-Color-Purity Red Fluorescence Emission through Edge Amino Protonation. *Nat. Commun.* **2021**, *12*, 6856. [CrossRef]
25. Jiang, K.; Sun, S.; Zhang, L.; Lu, Y.; Wu, A.; Cai, C.; Lin, H. Red, Green, and Blue Luminescence by Carbon Dots: Full-Color Emission Tuning and Multicolor Cellular Imaging. Angew. *Chem. Int. Ed.* **2015**, *54*, 5360–5363. [CrossRef]
26. Petit, T.; Puskar, L.; Dolenko, T.; Choudhury, S.; Ritter, E.; Burikov, S.; Laptinskiy, K.; Brzustowski, Q.; Schade, U.; Yuzawa, H.; et al. Unusual Water Hydrogen Bond Network around Hydrogenated Nanodiamonds. *J. Phys. Chem. C* **2017**, *121*, 5185–5194. [CrossRef]
27. Prabhakar, N.; Näreoja, T.; von Haartman, E.; Şen Karaman, D.; Burikov, S.A.; Dolenko, T.A.; Deguchi, T.; Mamaeva, V.; Hänninen, P.E.; Vlasov, I.I.; et al. Functionalization of Graphene Oxide Nanostructures Improves Photoluminescence and Facilitates Their Use as Optical Probes in Preclinical Imaging. *Nanoscale* **2015**, *7*, 10410–10420. [CrossRef]
28. Liu, Q.; Guo, B.; Rao, Z.; Zhang, B.; Gong, J.R. Strong Two-Photon-Induced Fluorescence from Photostable, Biocompatible Nitrogen-Doped Graphene Quantum Dots for Cellular and Deep-Tissue Imaging. *Nano Lett.* **2013**, *13*, 2436–2441. [CrossRef]
29. Yin, B.; Deng, J.; Peng, X.; Long, Q.; Zhao, J.; Lu, Q.; Chen, Q.; Li, H.; Tang, H.; Zhang, Y.; et al. Green Synthesis of Carbon Dots with Down- and up-Conversion Fluorescent Properties for Sensitive Detection of Hypochlorite with a Dual-Readout Assay. *Analyst* **2013**, *138*, 6551. [CrossRef]
30. Zhu, A.; Luo, Z.; Ding, C.; Li, B.; Zhou, S.; Wang, R.; Tian, Y. A Two-Photon "Turn-on" Fluorescent Probe Based on Carbon Nanodots for Imaging and Selective Biosensing of Hydrogen Sulfide in Live Cells and Tissues. *Analyst* **2014**, *139*, 1945–1952. [CrossRef]
31. Ha, H.D.; Jang, M.-H.; Liu, F.; Cho, Y.-H.; Seo, T.S. Upconversion Photoluminescent Metal Ion Sensors via Two Photon Absorption in Graphene Oxide Quantum Dots. *Carbon* **2015**, *81*, 367–375. [CrossRef]

32. Ni, Y.; Han, Z.; Ren, J.; Wang, Z.; Zhang, W.; Xie, Z.; Shao, Y.; Zhou, S. Ultralow Threshold Lasing from Carbon Dot–Ormosil Gel Hybrid-Based Planar Microcavity. *Nanomaterials* **2021**, *11*, 1762. [CrossRef] [PubMed]
33. Huang, M.; Liang, X.; Zhang, Z.; Wang, J.; Fei, Y.; Ma, J.; Qu, S.; Mi, L. Carbon Dots for Intracellular PH Sensing with Fluorescence Lifetime Imaging Microscopy. *Nanomaterials* **2020**, *10*, 604. [CrossRef] [PubMed]
34. Ren, J.; Weber, F.; Weigert, F.; Wang, Y.; Choudhury, S.; Xiao, J.; Lauermann, I.; Resch-Genger, U.; Bande, A.; Petit, T. Influence of Surface Chemistry on Optical, Chemical and Electronic Properties of Blue Luminescent Carbon Dots. *Nanoscale* **2019**, *11*, 2056–2064. [CrossRef]
35. Ye, X.; Xiang, Y.; Wang, Q.; Li, Z.; Liu, Z. A Red Emissive Two-Photon Fluorescence Probe Based on Carbon Dots for Intracellular PH Detection. *Small* **2019**, *15*, 1901673. [CrossRef]
36. Ţucureanu, V.; Matei, A.; Avram, A.M. FTIR Spectroscopy for Carbon Family Study. *Crit. Rev. Anal. Chem.* **2016**, *46*, 502–520. [CrossRef]
37. Vervald, A.M.; Lachko, A.V.; Kudryavtsev, O.S.; Shenderova, O.A.; Kuznetsov, S.V.; Vlasov, I.I.; Dolenko, T.A. Surface Photoluminescence of Oxidized Nanodiamonds: Influence of Environment PH. *J. Phys. Chem. C* **2021**, *125*, 18247–18258. [CrossRef]
38. Fan, T.; Zeng, W.; Tang, W.; Yuan, C.; Tong, S.; Cai, K.; Liu, Y.; Huang, W.; Min, Y.; Epstein, A.J. Controllable Size-Selective Method to Prepare Graphene Quantum Dots from Graphene Oxide. *Nanoscale Res. Lett.* **2015**, *10*, 55. [CrossRef]
39. Lee, D.W.; De Los Santos, V.L.; Seo, J.W.; Felix, L.L.; Bustamante, D.A.; Cole, J.M.; Barnes, C.H.W. The Structure of Graphite Oxide: Investigation of Its Surface Chemical Groups. *J. Phys. Chem. B* **2010**, *114*, 5723–5728. [CrossRef]
40. Zhang, H.; Huang, H.; Ming, H.; Li, H.; Zhang, L.; Liu, Y.; Kang, Z. Carbon Quantum Dots/Ag3PO4 Complex Photocatalysts with Enhanced Photocatalytic Activity and Stability under Visible Light. *J. Mater. Chem.* **2012**, *22*, 10501. [CrossRef]
41. Mintz, K.J.; Mercado, G.; Zhou, Y.; Ji, Y.; Hettiarachchi, S.D.; Liyanage, P.Y.; Pandey, R.R.; Chusuei, C.C.; Dallman, J.; Leblanc, R.M. Tryptophan Carbon Dots and Their Ability to Cross the Blood-Brain Barrier. *Colloids Surf. B Biointerfaces* **2019**, *176*, 488–493. [CrossRef] [PubMed]
42. De, B.; Karak, N. A Green and Facile Approach for the Synthesis of Water Soluble Fluorescent Carbon Dots from Banana Juice. *RSC Adv.* **2013**, *3*, 8286. [CrossRef]
43. Emam, A.N.; Loutfy, S.A.; Mostafa, A.A.; Awad, H.; Mohamed, M.B. Cyto-Toxicity, Biocompatibility and Cellular Response of Carbon Dots–Plasmonic Based Nano-Hybrids for Bioimaging. *RSC Adv.* **2017**, *7*, 23502–23514. [CrossRef]
44. Khmeleva, M.Y.; Laptinskiy, K.A.; Kasiyanova, P.S.; Tomskaya, A.E.; Dolenko, T.A. Dependence of the photoluminescence of carbon dots with different functionalization of the surface on the water pH. *Opt. Spectrosc.* **2022**, *6*, 882–889. (In Russian) [CrossRef]

applied sciences

Article

Terahertz Photoconductive Antenna Based on a Topological Insulator Nanofilm

Kirill A. Kuznetsov [1], Daniil A. Safronenkov [1], Petr I. Kuznetsov [2] and Galiya Kh. Kitaeva [1,*]

[1] Faculty of Physics, Lomonosov Moscow State University, 119991 Moscow, Russia; kirill-spdc@yandex.ru (K.A.K.); safronenkov.da14@physics.msu.ru (D.A.S.)
[2] Kotelnikov IRE RAS (Fryazino branch), 141190 Fryazino, Russia; pik218@fireras.su
* Correspondence: gkitaeva@physics.msu.ru

Abstract: In this study, the efficient generation of terahertz radiation by a dipole photoconductive antenna, based on a thin island film of a topological insulator, was experimentally demonstrated. The performance of the $Bi_{1.9}Sb_{0.1}Te_2Se$ antenna was shown to be no worse than those of a semiconductor photoconductive antenna, which is an order of magnitude thicker. The current–voltage characteristics were studied for the photo and dark currents in $Bi_{1.9}Sb_{0.1}Te_2Se$. The possible mechanisms for generating terahertz waves were analyzed by comparing the characteristics of terahertz radiation of an electrically biased and unbiased topological insulator.

Keywords: topological insulator; photoconductive antenna; terahertz waves

Citation: Kuznetsov, K.A.; Safronenkov, D.A.; Kuznetsov, P.I.; Kitaeva, G.K.. Terahertz Photoconductive Antenna Based on a Topological Insulator Nanofilm. *Appl. Sci.* **2021**, *11*, 5580. https://doi.org/10.3390/app11125580

Academic Editor: Gennady M. Mikheev

Received: 14 May 2021
Accepted: 15 June 2021
Published: 16 June 2021

1. Introduction

Currently, topological insulators (TIs) are considered to be promising new 2D materials for terahertz (THz) generators and detectors [1–3]. TIs are characterized by the presence of stable (topologically protected against backscattering), conducting surface states of an electronic gas. TIs can be used in the development of various interdisciplinary connections in many modern areas, including spintronics [4], laser physics [5], and quantum information [6].

Angle-resolved photoemission spectroscopy (ARPES) detects the presence of Dirac electrons with nondegenerate spins in many solids [7]. In addition to ARPES, there are other methods that are sensitive to edge electronic states. In [8], a new method was proposed for detecting the surface states of electrons based on the photo-electromagnetic effect (PEM), which is not sensitive to bulk conductivity. In [9], ultrafast manipulation of topologically enhanced surface transport was demonstrated using THz and mid-IR pulses in bismuth selenide. Correspondingly, the generation of THz radiation can also be sensitive to surface electron transfer. Hamh et al. demonstrated a circular anisotropy of the photon drag effect (PDE) in Bi_2Se_3, which proves optical coupling with topological surface states [10].

In a number of previous studies, the generation of THz radiation under the action of pulsed optical pumping was investigated in bismuth chalcogenides of double [11,12] and quaternary [13] chemical compositions. A comprehensive review of the photovoltaic effects in Bi_2Te_3 and Sb_2Te_3 was given in [14]. A number of conclusions were drawn regarding the prevailing mechanisms of the generation of THz radiation, due to the excitation of photocurrents in the epitaxial films of topological insulators in the absence of any external electric bias field. In [15], it was shown that the contribution to the photocurrent from the PDE is an order of magnitude greater than the contribution from the diffusion current, and two times higher than the contribution from the drift current.

Topological insulators of quaternary compositions $Bi_{2-x}Sb_xTe_{3-y}Se_y$ are actively studied in connection with the possibility of smooth changes in their properties over a wide range. In [16], it was shown that there is a certain optimal curve (the "Ren's curve") in the composition-structure diagram y(x), where the properties of electronic surface states are most pronounced. This is associated with a significant suppression of the bulk contribution to the conductivity, as the actions of acceptors and donors cancel each other, which leads to the predominance of surface electron transport. $Bi_{2-x}Sb_xTe_{3-y}Se_y$ films with compositions located under the Ren's curve, as a rule, predominantly have p-type conductivity; while compositions located above this curve have conductivity of the n-type.

At present, research is continuing on the most efficient methods for generating and detecting terahertz radiation. Photoconductive semiconductor antennas are effective devices in the THz range for applications in spectroscopy, and various imaging and data transmission systems [17,18]. Due to the availability of components in the telecommunications industry, modern systems often use optical excitation at a wavelength of 1550 nm with femtosecond pulses of an erbium fiber laser [19,20]. Semiconductor structures, based on indium-gallium arsenide, were well studied as photoconductive materials for THz due to their suitable optical absorption wavelength of 1.5 μm. Recently, THz detectors fabricated from rhodium (Rh)-doped $In_{0.53}Ga_{0.47}As$ showed a record peak dynamic range of 105 dB and a bandwidth up to 6.5 THz [21]. A significant improvement in the characteristics of these devices was achieved with the help of such nanotechnological tools as plasmonic contact gratings, arrays of optical nanoantennas, optical nanocavities [22,23], and non-standard orientation of the substrates [24,25].

Another important area is the development and optimization of couplers at the output surfaces of photoconductive antennas. The performance of planar terahertz antennas can be improved by using hemispherical dielectric lenses. However, this increases the weight and volume of the antenna. Artificially created composite metamaterials have recently attracted considerable attention from both the scientific and engineering communities due to their unconventional properties [26–30].

It is also of interest to use various other photoconductive materials for their application in antennas pumped by telecom wavelength lasers. Singh et al. [31] presented a significantly ultra-broadband (extending up to 70 THz) THz emission from an Au-implanted Ge emitter that was compatible with mode-locked fiber lasers operating at wavelengths of 1.1 and 1.55 μm.

The development of new photoconductive nanomaterials for THz PCAs still remains a relevant and urgent task. In our previous work [32], we demonstrated amplification of the THz radiation power by applying a weak external electric field to a TI. In this work, we study the generation of a photoconductive antenna under the application of a strong bias external field.

2. Proposed System Design

A TI sample of $Bi_{1.9}Sb_{0.1}Te_2Se$ (BSTS) was prepared with a composition near the Ren's curve at the composition-structure diagram. It was grown by metalorganic vapor-phase epitaxy (MOVPE) on a (0001) sapphire substrate with a 10-nm zinc telluride (ZnTe) buffer layer of orientation (111) at the atmospheric pressure of hydrogen in a horizontal quartz reactor. The sources of organometallic compounds of bismuth, antimony, zinc, tellurium, selenium, trimethyl bismuth, trimethyl antimony, diethyl zinc, diethyl tellurium, and diethyl selenium were used. The ZnTe buffer layer was grown in one technological cycle with a BSTS film at a temperature of 463 °C. To determine the elemental composition of the films, we used X-MaxN energy dispersive X-ray spectrometer with an electron microscope. A thin film of BSTS, 40 nm thick and of necessary composition, was selected for THz generation (Figure 1a).

A dipole antenna was fabricated by patterning Ti/Pd/Au metal electrodes (of corresponding thicknesses 50/120/200 nm) onto BSTS film. The gap between electrodes was 20 μm. Templates were preliminarily applied using the photolithography method. The fabricated TI-antenna had an impedance of 490 Ω. A hemispherical high-resistance silicon lens of 1 cm diameter was mounted on the surface of the Al_2O_3 substrate to collect the generated THz radiation (Figure 1c).

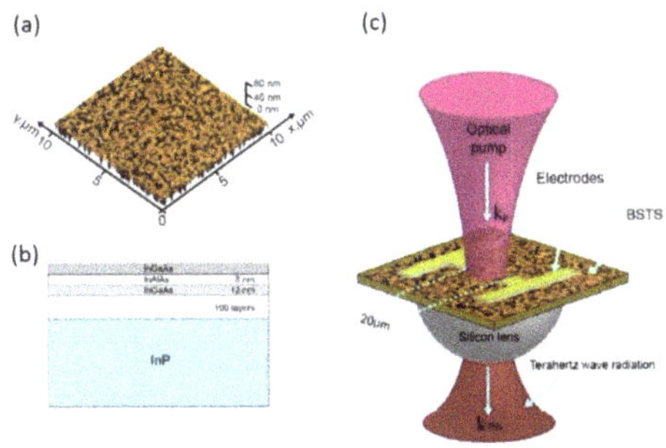

Figure 1. The (**a**) atomic force microscopy image of BSTS sample surface; (**b**) sketch of the InGaAs/InAlAs heterostructure; and (**c**) sketch of the antenna emitter.

To analyse the generation efficiency, we compared the generated THz pulses from two photoconductive antennas with the same dipole structure, one fabricated on the BSTS film (TI PCA) and the other on a semiconductor multilayer heterostructure (semiconductor PCA). Another dipole photoconductive antenna, fabricated on a low-temperature grown $In_{0.5}Ga_{0.5}As/In_{0.5}Al_{0.5}As$ superlattice, was used as a second antenna for reference. The parameters of similar spiral antenna were previously studied and reported in [33]. The multilayer heterostructure InGaAs/InAlAs for semiconductor PCA was grown on an InP (111) substrate (Figure 1b). Such structures are usually prepared by molecular beam epitaxy and exhibit a high conversion efficiency of optical radiation into the broadband THz radiation. The overall thickness of the InGaAs/InAlAs heterostructure was 2000 nm, and the total thickness of all InGaAs layers was 1200 nm. An impedance semiconductor PCA was 12.8 kΩ.

We measured the room temperature absorption coefficient at the pump wavelength. It was 4300 cm^{-1} for our InGaAs layer lattice and 120,000 cm^{-1} for the BSTS sample. In both cases, the thickness of the antenna material was less than the pump field penetration depth. Thus, the selected film thicknesses, 40 nm in the case of BSTS and 1200 nm of the total thickness of the InGaAs layers, provided almost uniform pumping illumination throughout the depth of the samples.

Figure 2a shows a photograph of the board. The electrodes were deposited on the samples of photoconductive films with five dipole gaps (Figure 2b). One of the five inter-electrode gaps, located above the most homogeneous region in the case of the topological insulator nanofilm, was chosen. Then, by heating with hot air from a soldering station, the electrodes were soldered to the board. A photograph of electrodes under a microscope is shown in Figure 2c. A hemispherical high-resistivity silicon lens 1 cm in diameter was pressed onto the substrate from above.

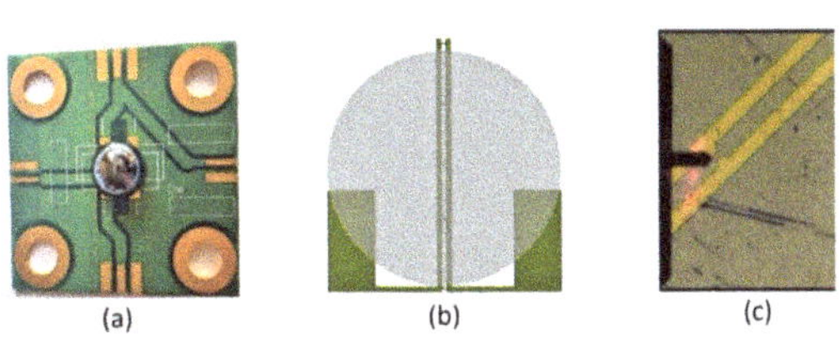

Figure 2. A (**a**) photo of the circuit board assembly; (**b**) sketch of the electrodes, with the gray circle representing the projection of the silicon lens onto the plane of the electrodes; and (**c**) image of the inter-electrode gap under a microscope. The gap width is 20 μm.

The measurements of THz generation characteristics were conducted on an experimental setup schematically presented in Figure 3. An Er^{3+}fiber laser was used as a source for optical pulsed pumping at a wavelength of 1560 nm, pulse repetition rate of 70 MHz, and pulse duration of 100 fs. Eighty percent of the radiation power (approximately 100 mW in the average intensity) was directed onto the antenna under study, and the last 20% of the radiation was used to run a commercial (Menlo Systems) semiconductor heterostructure photoconductive antenna. The latter was used for the detection of the waveforms of emitted THz fields. The pump beam was focused on the TI PCA1 with a short-focus (5 mm) lens. The parabolic mirrors collected the THz radiation and focused it on the input silicon lens of the antenna-detector PCA2.

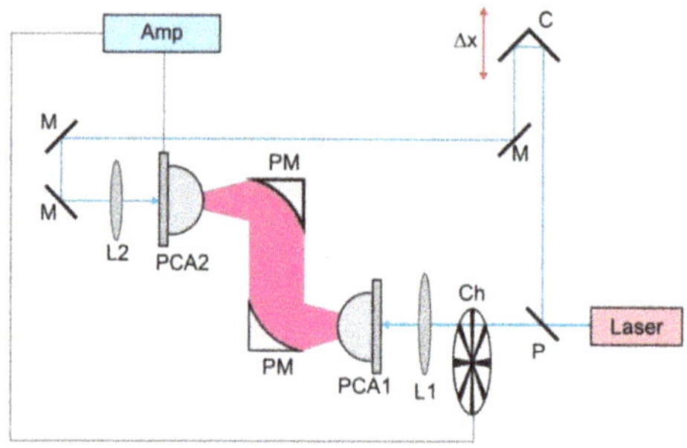

Figure 3. Schematics of the terahertz time-domain spectroscopy setup.

The photocurrents from the PCAs' were measured by a micro-ammeter and by a lock-in-amplifier at different bias voltages and laser pump powers. Sufficiently high photocurrents were measured simultaneously by using a current-sensitive input of the lock-in-amplifier. After the calibration of the lock-in-amplifier output data in microamperes, this sensitive device was used for measuring the lower photocurrents against a background of significant dark currents.

3. Results

A waveform of THz radiation, recorded from TI PCA at a bias voltage of 20 V, and its spectrum, obtained using fast Fourier transform, are shown by red curves in Figure 4a,b, respectively. The spectra in Figure 4b are "bumpy" due to re-reflection in the substrates and absorption by the water vapor. They extend up to 2 THz, which corresponds to the spectral range of the antenna-detector. The dynamic range of the obtained TI PCA radiation was approximately 60 dB.

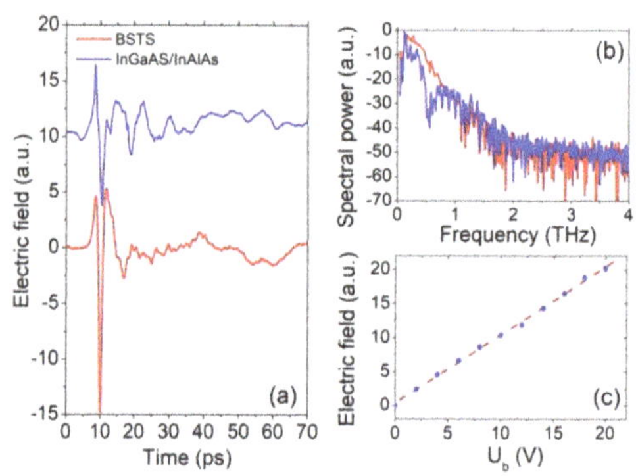

Figure 4. The (**a**) time-domain waveforms of dipole antennas based on BSTS (red) and InGaAs/InAlAs (blue); (**b**) fast Fourier transform spectra of BSTS (red) and InGaAs/InAlAs (blue) antennas; and (**c**) THz electric field amplitude versus bias voltage for BSTS antenna.

The blue curve in Figure 4a shows a waveform of THz radiation from a similar dipole semiconductor PCA biased at the same voltage U = 20 V (the curves were vertically shifted artificially). Though the thicknesses of active media differed by almost an order of magnitude (40 nm and 2000 nm), the amplitudes of the generated THz fields were comparable to each other. Even the peak-to-peak amplitude was 1.5 times more in case of thin TI PCA. Hence, we can conclude that, with the unit thickness of the medium, generation using BSTS is much more efficient than that using InGaAs/InAlAs. Figure 4c shows the dependence of the electric field amplitude on the bias voltage for TI PCA. The linear character of this dependence is clearly seen in the wide voltage range. At higher voltages, the THz signal was not recorded to avoid possible damage to the sample due to electrical breakdown.

As the generated photocurrents are responsible for the generation of THz radiation, it was interesting to measure them in the range of acting bias voltages of 0–20 V. First, the steady current-voltage characteristic was recorded, i.e., the dependence of the dark current on the voltage in the absence of incident optical pumping (Figure 5a). This characteristic was well approximated up to 13 V by a linear curve. However, at higher voltages some deviation from the linear fit was observed, which can be explained by a decrease in the resistance of the sample due to heating. We recorded the heating of the BSTS from room temperature to 50 °C by a thermocouple at a maximal bias voltage, U_b = 20 V. Figure 5b demonstrates the dependence of the average photocurrent on the bias voltage. It was also well described by a linear law over the entire range of U_b. For comparison, a similar dependence was shown for the semiconductor antenna, which lies well below the curve for the TI PCA. Figure 5c shows the obtained dependence of the TI PCA photocurrent on the pump power. It is linear; this is a typical situation for photoconductive antennas operating far from breakdown voltages and carrier screening effects.

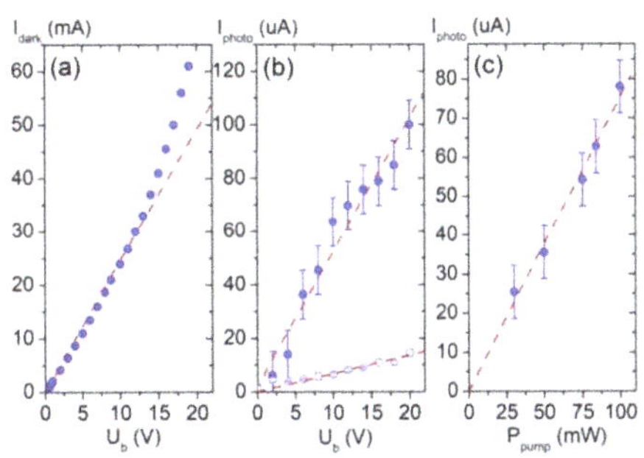

Figure 5. The (**a**) I-V characteristic of TI PCA measured without pump radiation; (**b**) filled circles: voltage dependence of the time-averaged photo-excited current of TI PCA; open circles: the same dependence for InGaAs/InAlAs PCA; and (**c**) dependence of the time-averaged photo-excited current on the pump power for TI PCA.

Figure 6a demonstrates the phase change of the generated THz radiation by 180° when the polarity of the bias voltage U_b =15 V was reversed. If we summarize the two signals obtained at inverse voltages, we realise practically the same waveform as in the absence of an external voltage (Figure 6b). Similar results were observed at a higher voltage of 20 V.

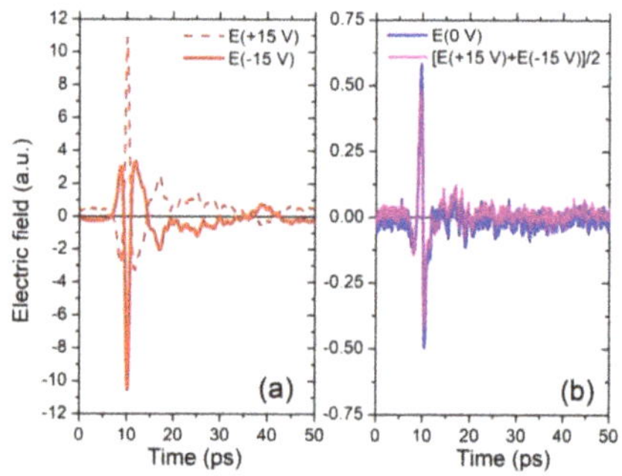

Figure 6. The (**a**) THz pulses at opposite bias voltages; and (**b**) comparison of the unbiased THz pulse (blue) with the result of averaging of two waveforms from oppositely biased antennas (magenta).

4. Discussion

In general, a fairly large set of different mechanisms can be used to explain THz generation in TIs of double and quaternary compositions. The most instantaneous one is direct optical rectification of the laser pump field [11] due to nonlinear polarization.

$$\mathbf{P}(t) \sim \int_{-\infty}^{t} \hat{\chi}^{(2)}(t-t', t-t'') : \mathbf{E}(t')\mathbf{E}^{*}(t'')dt'dt'' + c.c. \tag{1}$$

Here, $\mathbf{E}(t)$ stands for the laser pump field and $\hat{\chi}^{(2)}$ describes the second-order nonlinear-optical susceptibility. Due to the centrosymmetric structure of TI volume, the components of $\hat{\chi}^{(2)}$ tensor are non-zero only at its surface and, thus, this effect is related specifically to surface excitations. However, its contribution was much smaller [34] than the THz field generated by transient currents of photo-induced charge carriers. The current-induced THz field depends on time derivatives of the two types of photocurrents, usually referred as linear ($\mathbf{j}_L$) and nonlinear ($\mathbf{j}_{NL}$) ones [15]:

$$\mathbf{E}_{THz}(t) \sim \frac{\partial \mathbf{j}_L(t)}{\partial t} + \frac{\partial \mathbf{j}_{NL}(t)}{\partial t}. \tag{2}$$

In the frame of the simplest classical description, both types of currents should depend on temporal behaviour of the carrier's concentration $N(t)$, mobility μ, and some specific field $\tilde{\mathbf{E}}$ as:

$$\mathbf{j}_{L,NL}(t) = e\mu N(t)\tilde{\mathbf{E}}. \tag{3}$$

In our case of above-bandgap photo-excitation, the amplitude of the $N(t)$ pulse scaled up with the pump pulse energy. As was reasonably argued in [35], the contribution from photo-induced electrons to both types of currents in BSTS was considerably more than the contribution from holes. A presence of non-zero dc field is necessary for directional motion of charge carriers in this description, regardless of whether it is applied externally or originates due to specific internal processes. The origin of, and the character of its possible temporal variation, are different for different THz generation mechanisms.

In case of linear currents $\mathbf{j}_L$, the temporal variation of $\tilde{\mathbf{E}}$ was negligible. In a photoconductive antenna, $\tilde{\mathbf{E}} = \mathbf{E}_{dc}$ is created by an external bias voltage source. With significantly smaller magnitudes, $\tilde{\mathbf{E}}$ also exists as a surface depletion field or due to the photo-Dember effect [15], regardless of whether an additional lateral external field is applied. The surface depletion and photo-Dember fields are directed along the normal to the TI surface and determine the motion of volume carriers even in case of zero applied voltage. In addition, some surface built-in field can exist in the lateral direction and be responsible for the specific contribution of surface carriers to THz generation. For all generation effects caused by different-type linear currents, the THz waveforms should have almost the same shape, repeating the shape of $\partial N(t)/\partial t$. This shape is clearly presented in waveforms like those in Figure 6a. The waveforms were obtained in the case where the high external bias field made $\mathbf{j}_L$ more than other nonlinear contributions to the net photo-induced current. It was seen that changing the bias voltage sign did not cause any observable changes in the THz field magnitude, apart from reversing its sign. Moreover, both curves in Figure 6b demonstrate the zero residual THz field at delay times, when contribution of linear currents should be maximal. Thus, we concluded that the effects of lateral and longitudinal built-in fields were negligible at zero voltage, and were not linearly enhanced when the bias voltage was applied to the TI antenna.

By comparing the mean photocurrents detected in the TI antenna and the semiconductor PCA at the same bias voltage (Figure 5b), we obtained a ratio between the photocurrents (~1:7), which roughly corresponded to the ratio between the mobilities of bulk photoelectrons in both materials, $\mu_{semi} = 30\ \text{cm}^2/(\text{V}\cdot\text{s})$ in our semiconductor InGaAs/InAlAs superlattice [33], and $\mu_{TI} = 166\ \text{cm}^2/(\text{V}\cdot\text{s})$ in BSTS [36]. According to (3) this meant that the mean concentrations of photo-induced bulk electrons were the same in both materials. However, the peak-to-peak THz field amplitudes detected in TI and semiconductor PCAs (Figure 4a) refered to each other as 1.6:1. This indicates that the photocurrent change rate in BSTS was smaller than in the InGaAs/InAlAs superlattice. Thus, we estimated the ratio between the electron relaxation times in the both materials as $\tau_{TI}/\tau_{semi} \sim (E_{THz,semi}/E_{THz,TI})(\mu_{TI}/\mu_{semi}) \sim 3.5$. Taking the previously measured value of the characteristic relaxation time for the InGaAs/InAlAs superlattice $\tau_{semi} = 1.7$ ps, we obtained the relaxation time for the bulk photo-induced electrons in BSTS as approximately $\tau_{TI} \approx 6$ ps.

Let us consider which relaxation process this time describes. In the case of BSTS, with extremely low initial concentration of free bulk carriers, the THz generation process began with the absorption of pump photons and the formation of electrons in the conduction band (Figure 7). The characteristic time of this process was of the order of the pump duration t_0. Time t_1 was of the order of a few picoseconds and, within this interval, a relaxation of bulk electrons took place due to the phonon emission [37]. The second type of relaxation process was an interband transition from the bulk conduction band to the surface states at the Dirac cone. As the Fermi level E_f lies in the band of surface states (SSs) in a BSTS with a composition near the Ren's curve, SSs above E_f were primarily unoccupied. The characteristic time t_2 of this process is rather long, approximately 5–10 ps [37,38]. Finally, the relaxation of SSs was the third relaxation mechanism, with time t_3 as hundreds of femtoseconds or several picoseconds [9,39]. The direct interband transitions from the conduction band to the valence band are rather long and usually last from hundreds to thousands of picoseconds. These slow processes do not significantly contribute to generation at THz frequencies. Our estimate of the relaxation time corresponds to t_2. Thus, we concluded that the process of population of the edge states by photo-induced bulk electrons made up the main contribution to the formation of the THz generation spectrum of the BSTS TI PCA.

The nonlinear $\mathbf{j}_{NL}$ currents are known to appear in unbiased TIs due to photo-galvanic (PG) and photon-drag (PD) effects [10–15,40,41]. Currents induced in these processes should quadratically depend on the laser pump field, both under sub-bandgap excitation [40] and under above-bandgap [10,13] laser excitation. Phenomenologically, dependences of these currents on the laser electric field for the non-stationary case [42] are described as:

$$\mathbf{j}_{NL}^{PG}(t) = \int_{-\infty}^{t} \widehat{\beta}(t-t', t-t'') : \mathbf{E}(t')\mathbf{E}^*(t'')dt'dt'' + c.c., \tag{4}$$

$$\mathbf{j}_{NL}^{PD}(t) = \int_{-\infty}^{t} \widehat{\gamma}(t-t', t-t'') : \mathbf{E}(t')\frac{\partial \mathbf{E}^*(t'')}{\partial \mathbf{r}}dt'dt'' + c.c. \tag{5}$$

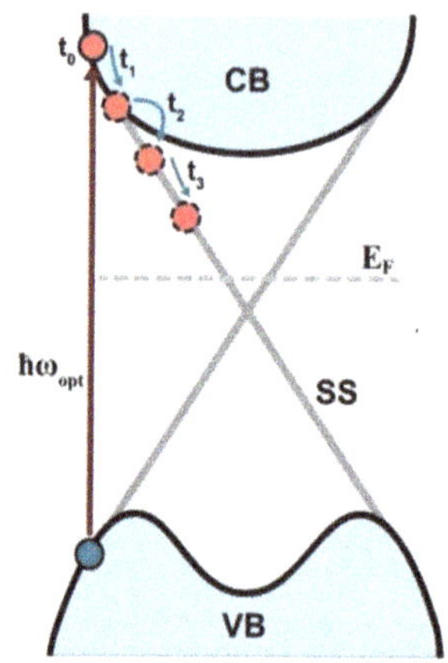

Figure 7. Relaxation of carriers in BSTS film.

The PG effect takes place due to third-order tensor $\widehat{\beta}(t-t', t-t'')$, which is non-zero only at the surface of TI. It is tied directly to the specific surface carriers, while both the bulk and surface electrons take part in PD effect via the fourth-order tensor $\widehat{\gamma}(t-t', t-t'')$. As it was shown in previous experimental works [40,41], in case of the normal incidence of the pump beam on BSTS, PD effect is negligible in comparison with PG generation. In our case, at zero bias voltage we detected the same bipolar shapes of the temporal waveforms (Figure 6b) as in [15]. Thus, we can relate the THz fields observed in the unbiased TI to the PG effect.

By comparing expression (4) for PG current with the general presentation of the induced currents in form (3), one can see that (4) seemingly depicts an effect of resonant nonlinear generation of an effective field $\tilde{\mathbf{E}}$, which is responsible for the directional motion of free carriers. More accurate calculations, performed by considering the kinetic Boltzmann distribution equation for the electron distribution function within the semi-classical approach [40], can be treated so that an effective field $\tilde{\mathbf{E}}$ appears here as a result of two processes: an alignment of carrier moments along the pump electric field, and an asymmetric scattering of carriers at the surface of TI [41]. However, this theory describes quadratic dependence on the pump field for THz field generated under below-bandgap excitation, but needs further assumptions to explain generation under above-bandgap pumping.

Indeed, the below-bandgap excitation did not change the carrier's concentration, $N(t)$ was almost constant in (3), and only the temporal change of effective $\tilde{\mathbf{E}}$ determined a short PG current pulse, which scaled up quadratically with the pump electric field. At the same time, when the pump photon energy exceeded the bandgap, the concentration of charge carriers also grew quadratically with the pump field. This was the case for bulk electrons, while a saturation of the concentration of the surface electrons may occur. However, no fourth-order dependence on the pump field was observed under above-bandgap excitation.

So, taking all the above considerations into account, we can conclude that: (1) the impact of the PG effect, and accompanying effective PG field $\tilde{\mathbf{E}}$, into THz generation in the biased THz antenna was negligible. Generation occurred due to photo-excited *bulk* carriers and its relaxation through the surface states in the presence of the constant field $\mathbf{E}_{dc}$. The temporal shape of $N(t)$ determined THz generation waveform and the width of the antenna spectral response; and (2) in the absence of an external biased field, THz generation took place due to the PG effect by *surface* electrons, which had almost the same concentration during the whole generation pulse. The temporal shape of effective $\tilde{\mathbf{E}}(t)$ determined the THz generation specific waveform and the width of the BSTS spectral response.

5. Conclusions

In conclusion, we demonstrated a working photoconductive antenna with a dynamic range of order of 60 dB based on the nanofilm of the TI $Bi_{2-x}Sb_xTe_{3-y}Se_y$. In addition to the known nonlinear effects, the photoconductive mechanism of THz generation was realized in TIs because of the relaxation of bulk electrons through the surface states. A comparison of photoconductive antennas based on TI and semiconductor heterostructure showed that, despite the great difference in the thicknesses by more than one order, the THz fields of both PCAs were comparable in amplitude. The time-averaged photocurrent and dark currents arising in the TI PCA were measured and their linear dependences on the bias voltage were demonstrated.

TI antennas represent promising tools for generating THz radiation. Due to being extremely thin, they can be easily integrated into nanophotonic devices. They are inexpensive to manufacture, as the growing of a TI film takes several minutes without the need for difficult growth conditions, such as those required for MBE-technology. Their conversion efficiency can be further increased by applying plasmon gratings. Lastly, TI antennas are able to withstand large dark currents without being destroyed.

Author Contributions: Conceptualization, K.A.K., and P.I.K.; methodology, G.K.K.; software, D.A.S.; validation, K.A.K., and D.A.S.; formal analysis, G.K.K.; investigation, K.A.K., D.A.S., and G.K.K.; sample growth and characterization, P.I.K.; writing—original draft preparation, K.A.K., and G.K.K.; writing—review and editing, K.A.K., G.K.K., and D.A.S.; visualization, D.A.S.; funding acquisition, K.A.K. All authors have read and agreed to the published version of the manuscript.

Funding: This research was funded by the Russian Foundation of Basic Research, grant numbers 18-29-20101 and 19-02-00598.

Institutional Review Board Statement: Not applicable.

Informed Consent Statement: Not applicable.

Acknowledgments: The authors are grateful to D.S. Ponomarev, A.N. Klochkov, D.V. Lavrukhin, V.B. Novikov, and S.N. Molotkov for helpful discussions.

Conflicts of Interest: The authors declare no conflict of interest.

References

1. Zhang, X.; Wang, J.; Zhang, S.-C. Topological insulators for high-performance terahertz to infrared applications. *Phys. Rev. B* **2010**, *82*, 245107. [CrossRef]
2. Viti, L.; Coquillat, D.; Politano, A.; Kokh, K.A.; Aliev, Z.S.; Babanly, M.B.; Tereshchenko, O.E.; Knap, W.; Chulkov, E.V.; Vitiello, M.S. Plasma-Wave Terahertz Detection Mediated by Topological Insulators Surface States. *Nano Lett.* **2016**, *16*, 80–87. [CrossRef]
3. West, D.; Zhang, S.B. Thin-film topological insulators for continuously tunable terahertz absorption. *Appl. Phys. Lett.* **2018**, *112*, 091601. [CrossRef]
4. Li, C.H.; van 't Erve, O.M.J.; Yan, C.; Li, L.; Jonker, B.T. Electrical detection of current generated spin in topological insulator surface states: Role of interface resistance. *Sci. Rep.* **2019**, *9*, 6906. [CrossRef] [PubMed]
5. Yan, P.; Lin, R.; Ruan, S.; Liu, A.; Chen, H.; Zheng, Y.; Chen, S.; Guo, C.; Hu, J. A practical topological insulator saturable absorber for mode-locked fiber laser. *Sci. Rep.* **2015**, *5*, 1–5. [CrossRef] [PubMed]
6. He, M.; Sun, H.; He, Q.L. Topological insulator: Spintronics and quantum computations. *Front. Phys.* **2019**, *14*, 43401. [CrossRef]
7. Ando, Y. Topological Insulator Materials. *J. Phys. Soc. Jpn.* **2013**, *82*, 102001. [CrossRef]
8. Egorova, S.G.; Chernichkin, V.I.; Ryabova, L.I.; Skipetrov, E.P.; Yashina, L.; Danilov, S.N.; Ganichev, S.D.; Khokhlov, D.R. Detection of highly conductive surface electron states in topological crystalline insulators $Pb_{1-x}Sn_xSe$ using laser terahertz radiation. *Sci. Rep.* **2015**, *5*, 11540. [CrossRef] [PubMed]
9. Luo, L.; Yang, X.; Liu, X.; Liu, Z.; Vaswani, C.; Cheng, D.; Mootz, M.; Zhao, X.; Yao, Y.; Wang, C.-Z.; et al. Ultrafast manipulation of topologically enhanced surface transport driven by mid-infrared and terahertz pulses in Bi_2Se_3. *Nat. Commun.* **2019**, *10*, 1–9. [CrossRef]
10. Hamh, S.Y.; Park, S.-H.; Jerng, S.-K.; Jeon, J.H.; Chun, S.-H.; Lee, J.S. Helicity-dependent photocurrent in a Bi_2Se_3 thin film probed by terahertz emission spectroscopy. *Phys. Rev. B* **2016**, *94*, 161405. [CrossRef]
11. Zhu, L.-G.; Kubera, B.; Mak, K.F.; Shan, J. Effect of Surface States on Terahertz Emission from the Bi_2Se_3 Surface. *Sci. Rep.* **2015**, *5*, 1–8. [CrossRef] [PubMed]

12. Tu, C.-M.; Chen, Y.-C.; Huang, P.; Chuang, P.-Y.; Lin, M.-Y.; Cheng, C.-M.; Lin, J.-Y.; Juang, J.-Y.; Wu, K.-H.; Huang, J.-C.A.; et al. Helicity-dependent terahertz emission spectroscopy of topological insulator Sb_2Te_3 thin films. *Phys. Rev. B* **2017**, *96*, 195407. [CrossRef]
13. Onishi, Y.; Ren, Z.; Novak, M.; Segawa, K.; Ando, Y.; Tanaka, K. Instantaneous Photon Drag Currents in Topological Insulators. *arXiv* **2014**, arXiv:1403.2492.
14. Plank, H.; Pernul, J.; Gebert, S.; Danilov, S.N.; König-Otto, J.; Winnerl, S.; Lanius, M.; Kampmeier, J.; Mussler, G.; Aguilera, I.; et al. Infrared/terahertz spectra of the photogalvanic effect in (Bi,Sb)Te based three-dimensional topological insulators. *Phys. Rev. Mater.* **2018**, *2*, 024202. [CrossRef]
15. Fang, Z.; Wang, H.; Wu, X.; Shan, S.; Wang, C.; Zhao, H.; Xia, C.; Nie, T.; Miao, J.; Zhang, C.; et al. Non-linear terahertz emission in the three-dimensional topological insulator Bi_2Te_3 by terahertz emission spectroscopy. *Appl. Phys. Lett.* **2019**, *115*, 191102. [CrossRef]
16. Ren, Z.; Taskin, A.A.; Sasaki, S.; Segawa, K.; Ando, Y. Optimizing $Bi_{2-x}Sb_xTe_{3-y}Se_y$ solid solutions to approach the intrinsic topological insulator regime. *Phys. Rev. B* **2011**, *84*, 165311. [CrossRef]
17. Castro-Camus, E.; Alfaro, M. Photoconductive devices for terahertz pulsed spectroscopy: A review. *Photon. Res.* **2016**, *4*, A36. [CrossRef]
18. Burford, N.M.; El-Shenawee, M.O. Review of terahertz photoconductive antenna technology. *Opt. Eng.* **2017**, *56*, 010901. [CrossRef]
19. Vieweg, N.; Rettich, F.; Deninger, A.; Roehle, H.; Dietz, R.; Göbel, T.; Schell, M. Terahertz-time domain spectrometer with 90 dB peak dynamic range. *J. Infrared Millim. Terahertz Waves* **2014**, *35*, 823–832. [CrossRef]
20. Globisch, B.; Dietz, R.J.B.; Kohlhaas, R.B.; Göbel, T.; Schell, M.; Alcer, D.; Semtsiv, M.; Masselink, W.T. Iron doped InGaAs: Competitive THz emitters and detectors fabricated from the same photoconductor. *J. Appl. Phys.* **2017**, *121*, 053102. [CrossRef]
21. Kohlhaas, R.B.; Breuer, S.; Nellen, S.; Liebermeister, L.; Schell, M.; Semtsiv, M.P.; Masselink, W.T.; Globisch, B. Photo-conductive terahertz detectors with 105 dB peak dynamic range made of rhodium doped InGaAs featured. *Appl. Phys. Lett.* **2019**, *114*, 221103. [CrossRef]
22. Lepeshov, S.; Gorodetsky, A.; Krasnok, A.; Rafailov, E.; Belov, P. Enhancement of terahertz photoconductive antenna operation by optical nanoantennas. *Laser Photonics Rev.* **2017**, *11*, 1600199. [CrossRef]
23. Yachmenev, A.E.; Lavrukhin, D.V.; Glinskiy, I.A.; Zenchenko, N.V.; Goncharov, Y.G.; Spektor, I.E.; Khabibullin, R.A.; Otsuji, T.; Ponomarev, D.S. Metallic and dielectric metasurfaces in photoconductive terahertz devices: A review. *Opt. Eng.* **2019**, *59*, 061608. [CrossRef]
24. Kuznetsov, K.A.; Galiev, G.B.; Kitaeva, G.K.; Kornienko, V.V.; Klimov, E.A.; Klochkov, A.N.; Leontyev, A.A.; Pushkarev, S.S.; Malrsev, P.P. Photoconductive antennas based on epitaxial films $In_{0.5}Ga_{0.5}As$ on GaAs (111)A and (100)A substrates with a metamorphic buffer. *Laser Phys. Lett.* **2018**, *15*, 076201. [CrossRef]
25. Galiev, G.B.; Grekhov, M.M.; Kitaeva, G.K.; Klimov, E.A.; Klochkov, A.; Kolentsova, O.S.; Kornienko, V.; Kuznetsov, K.A.; Maltsev, P.; Pushkarev, S. Terahertz-radiation generation in low-temperature InGaAs epitaxial films on (100) and (411) InP substrates. *Semiconductors* **2017**, *51*, 310–317. [CrossRef]
26. Zhang, Q.-L.; Si, L.-M.; Huang, Y.; Lv, X.; Zhu, W. Low-index-metamaterial for gain enhancement of planar terahertz antenna. *AIP Adv.* **2014**, *4*, 037103. [CrossRef]
27. Hussain, N.; Nguyen, T.K.; Han, H.; Park, I. Minimum Lens Size Supporting the Leaky-Wave Nature of Slit Dipole Antenna at Terahertz Frequency. *Int. J. Antennas Propag.* **2016**, *2016*, 1–8. [CrossRef]
28. Hussain, N.; Park, I. Design of a wide-gain-bandwidth metasurface antenna at terahertz frequency. *AIP Adv.* **2017**, *7*, 055313. [CrossRef]
29. Llombart, N.; Chattopadhyay, G.; Skalare, A.; Mehdi, I. Novel Terahertz Antenna Based on a Silicon Lens Fed by a Leaky Wave Enhanced Waveguide. *IEEE Trans. Antennas Propag.* **2011**, *59*, 2160–2168. [CrossRef]
30. Li, X.; Yin, W.; Khamas, S. An Efficient Photomixer Based Slot Fed Terahertz Dielectric Resonator Antenna. *Sensors* **2021**, *21*, 876. [CrossRef]
31. Singh, A.; Pashkin, A.; Winnerl, S.; Welsch, M.; Beckh, C.; Sulzer, P.; Leitenstorfer, A.; Helm, M.; Schneider, H. Up to 70 THz bandwidth from an implanted Ge photoconductive antenna excited by a femtosecond Er:fibre laser. *Light. Sci. Appl.* **2020**, *9*, 1–7. [CrossRef]
32. Kuznetsov, K.A.; Kitaeva, G.K.; Kuznetsov, P.I.; Yakushcheva, G.G. Generation of terahertz radiation from the island films of topological insulator $Bi_{2-x}Sb_xTe_{3-y}Se_y$. *AIP Adv.* **2019**, *9*, 015310. [CrossRef]
33. Kuznetsov, K.; Klochkov, A.; Leontyev, A.; Klimov, E.; Pushkarev, S.; Galiev, G.; Kitaeva, G. Improved InGaAs and InGaAs/InAlAs Photoconductive Antennas Based on (111)-Oriented Substrates. *Electronics* **2020**, *9*, 495. [CrossRef]
34. Braun, L.; Mussler, G.; Hruban, A.; Konczykowski, M.; Schumann, T.; Wolf, M.; Münzenberg, M.; Perfetti, L.; Kampfrath, T. Ultrafast photocurrents at the surface of the three-dimensional topological insulator Bi_2Se_3. *Nat. Commun.* **2016**, *7*, 13259. [CrossRef]
35. Park, S.H.; Hamh, S.Y.; Park, J.; Kim, J.S.; Lee, J.S. Possible flat band bending of the $Bi_{1.5}Sb_{0.5}Te_{1.7}Se_{1.3}$ crystal cleaved in an ambient air probed by terahertz emission spectroscopy. *Sci. Rep.* **2016**, *6*, 36343. [CrossRef]
36. Kuznetsov, K.A.; Kuznetsov, P.I.; Frolov, A.D.; Kovalev, S.P.; Ilyakov, I.E.; Ezhov, A.A.; Kitaeva, G.K. Bulk and surface terahertz conductivity of $Bi_{2-x}Sb_xTe_{3-y}Se_y$ topological insulators. *Opt. Eng.* **2021**, *60*, 082012. [CrossRef]

37. Lorenc, M.; Onishi, Y.; Ren, Z.; Segawa, W.; Kaszub, W.; Ando, Y.; Tanaka, K. Ultrafast carrier relaxation through Auger recombination in the topological insulator$Bi_{1.5}Sb_{0.5}Te_{1.7}Se_{1.3}$. *Phys. Rev. B* **2015**, *91*, 085306. [CrossRef]
38. Sobota, J.A.; Yang, S.; Analytis, J.G.; Chen, Y.L.; Fisher, I.R.; Kirchmann, P.S.; Shen, Z.-X. Ultrafast Optical Excitation of a Persistent Surface-State Population in the Topological Insulator Bi_2Se_3. *Phys. Rev. Lett.* **2012**, *108*, 117403. [CrossRef]
39. Kovalev, S.; Tielrooij, K.-J.; Deinert, J.-C.; Ilyakov, I.; Awari, N.; Chen, M.; Ponomaryov, A.; Bawatna, M.; de Oliveira, T.V.A.G.; Eng, L.M.; et al. Terahertz signatures of ultrafast Dirac fermion relaxation at the surface of topological insulators at room temperature. *arXiv* **2006**, arXiv:2006.03948.
40. Plank, H.; Ganichev, S.D. A review on terahertz photogalvanic spectroscopy of Bi_2Te_3- and Sb_2Te_3-based three dimensional topological insulators. *Solid State Electron.* **2018**, *147*, 44–50. [CrossRef]
41. Plank, H.; Golub, L.E.; Bauer, S.; Bel'kov, V.V.; Herrmann, T.; Olbrich, P.; Eschbach, M.; Plucinski, L.; Schneider, C.M.; Kampmeier, J.; et al. Photon drag effect in$(Bi_{1-x}Sb_x)_2Te_3$ three-dimensional topological insulators. *Phys. Rev. B* **2016**, *93*, 125434. [CrossRef]
42. Il'inskii, Y.A.; Keldysh, L.V. General Theory of Interaction of Electromagnetic Fields with Matter. In *Electromagnetic Response of Material Media*; Springer: Boston, MA, USA, 1994.

applied sciences

MDPI

Communication

Tiling Photonic Topological Insulator for Laser Applications

Petr N. Kim [1,2], Dmitry P. Fedchenko [1,2], Natalya V. Rudakova [1,2] and Ivan V. Timofeev [1,2,*]

1 Kirensky Institute of Physics, Federal Research Center KSC SB RAS, Krasnoyarsk 660036, Russia
2 Siberian Federal University, Krasnoyarsk 660041, Russia
* Correspondence: tiv@iph.krasn.ru

Featured Application: A new class of active topological photonic devices and laser arrays.

Abstract: A photonic topological insulator is a structure that isolates radiation in the bulk rather than at the edge (surface). Paradoxically, applications of such an insulator focus on its conducting edge states, which are robust against structural defects. We suggest a tiling photonic topological insulator constructed from identical prism resonators connected to each other. The light beam circulates inside the tiling bulk without propagation. However, we experimentally demonstrate a topologically-protected propagating state due to the disconnected faces of the edge resonators. The investigated state is robust against removing or attaching prism resonators. Moreover, the protection principle is phase-free and therefore highly scalable both in wavelength and resonator size. The tiling is suggested for active topological photonic devices and laser arrays.

Keywords: photonic topological insulator; total internal reflection; topological insulator laser

Citation: Kim, P.N.; Fedchenko, D.P.; Rudakova, N.V.; Timofeev, I.V. Tiling Photonic Topological Insulator for Laser Applications. *Appl. Sci.* **2023**, *13*, 4004. https://doi.org/10.3390/app13064004

Academic Editor: Vladimir M. Fomin

Received: 22 February 2023
Revised: 16 March 2023
Accepted: 20 March 2023
Published: 21 March 2023

1. Introduction

Recently, a topological insulator laser was demonstrated [1,2] which exhibits scatter-free edge-state transport of light in the laser cavity, immune to perturbations and disorder. The topological feature forces injection locking of many laser emitters to act as a single coherent laser. The concept led to a variety of configurations: electrical pumping of a quantum cascade laser with valley modes [3]; vertical emission [4]; and bulk and vortex lasing [5,6]. Simultaneously, one-dimensional active resonator arrays were described by a Su–Schrieffer–Heeger model [7], and then implemented and optimized for robust single-mode high-power lasers [8–10].

Topological invariance of a physical process means that it is described by some highly robust conserved quantity that does not change under continuous deformations in the parametric space [11]. For example, the number of holes of a connected smooth compact surface does not change as the surface deforms, without cutting or gluing. In particular, Emmy Noether's theorem connects symmetries to the existence of the corresponding conserved quantities, such as energy, momentum, angular momentum, electric charge, and others. Parametric deformations can break the symmetry, still preserving the topological invariant. Almost forty years ago, a topological invariant was introduced for a single-particle electron state in a combined periodic potential and out-of-plane magnetic field [12,13]. The similar wave-like behavior is inherent to the photon state in the periodic potential of the photonic crystal [14]. As the Bloch wavenumber adiabatically passes the loop trajectory within the Brillouin zone of the crystal, its wave function acquires some topological phase that defines the topological invariant of the photonic band, separated by an insulating bandgap from neighboring bands. According to the bulk-edge correspondence principle [15], the number of edge states in the bandgap is equal to the difference of topological invariants in neighboring bands. Here, one assumes the *spatial edge* of the array rather than frequency band edge. As usual, in the optical frequency range, the magnetic response is weak. Instead, a spin–orbit coupling [16], dynamical modulation [17], or gain-loss materials [2] can also give rise to nonzero topological phases separated by an insulating bandgap. This idea was easily transferred from photonic crystal to waveguides, including ring resonator platforms [18].

Quartz prism resonators serve well instead of ring resonators, as the excitation can propagate between prisms by the *frustrated total internal reflection* mechanism. For the latter structure, additional edge solutions arise at different incident angles [19]. Nonetheless, there is the question of realizing such a complex coupling utilizing synthetic dimensions [20], or in the microring fashion [21,22]. If the structure period is comparable to radiation wavelength, then the description engages the topological phase. Otherwise, when the resonators are much larger than the wavelength, a phase-free explanation of the edge state is appropriate in terms of the *edge trajectory* of the beam [19].

In this paper, we suggest a *tiling photonic topological insulator* (TPhTI) constructed from identical prism resonators. Compared to [19], this new design suggests a continuous connection of neighboring resonators without spatial gaps. The light beam circulates within a few prisms inside the tiling bulk, penetrating through the mutual faces of the prisms or turning the propagation direction by total internal reflection at the open faces. However, a topologically-propagating state occurs at the edge of the array due to the disconnected faces of edge prisms. This propagating state persists against removing or attaching new prism resonators.

Furthermore, we fabricated the TPhTI and demonstrated the topologically-protected propagating state experimentally. We also speculated on laser applications of the TPhTI containing the gain material.

2. Materials and Methods

To fabricate the TPhTI we used a detailed concept similar to [19]. Rectangular prism resonators (Figure 1a) were manufactured from quartz glass with a refractive index of 1.43. Their linear dimensions of $12 \times 12 \times 8$ mm^3 were ensured by the Maksutov method [23] with 10 um accuracy. To avoid accuracy problems with the half-wavelength gap and multi-layer covering, we chose the zero gap. The immersion liquid infiltrated the unintentional air gaps between the prism resonators to suppress the parasitic reflection and to ensure the lossless beam penetration between the resonators. Therefore, a liquid with a refractive index close to the refractive index of quartz glass was selected.

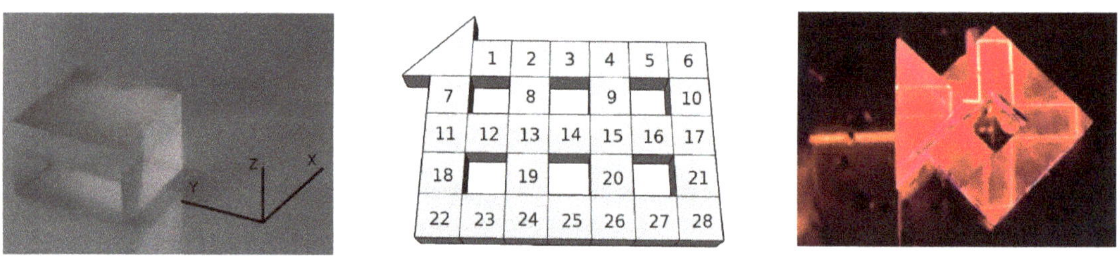

(a) (b) (c)

Figure 1. (**a**) The prism resonator made of quartz glass with a refractive index of 1.43. (**b**) A model of the tiling photonic topological insulator (TPhTI) composed of 28 prism resonators. (**c**) The TPhTI, composed of seven prism resonators.

The prism resonator tiling was precisely positioned on a flat platform, 200×200 mm^2 in size. A rectangular triangular prism was used as a coupler to lead the light beam into and out of the quartz glass at the angle of total internal reflection. The angle of total internal reflection is 44.37° according to Snell's law. The resonators were fixed on an adhesive composition of a mixture of vinyl acetate (30%) and acetone (70%) copolymers. The immersion liquid was an aqueous solution of glycerin with a density of 1.228 g per cubic centimeter. The refractive index of the immersion liquid was 1.45. No more than 0.01 mL of immersion liquid covered each face to avoid a meniscus on adjacent vertical faces, producing scattering and deviation of the beam. Before applying the immersion liquid, each prism was cleaned of dust and stains using crepe paper impregnated with a solution of water (80%), isopropanol (10%), metaxypropanol (5%), and nonionic surfactant (5%). Then, rectangular quartz prisms were connected, strictly in the order of numbering, as denoted in Figure 1b. After connecting each new segment, the operability of the tiling was checked using violet (405 nm) and red (650 nm) lasers (Figure 1c). Each prism was adjusted according to three degrees of freedom, namely one rotation and two translations along the axes X and Y (Figure 1a). The glass platform with assembled tiling and a triangular coupler was fixed on an optical bench next to the laser source.

3. Results

Figure 2 schematically demonstrates the stability of the beam trajectory; resonator #3, as denoted in Figure 1b, was removed in order to experimentally create the lattice defect of the TPhTI. Here, the green color indicates the trajectory of the beam. *In* and *Out* denote the entry and exit points of the beam, respectively. The input and output can be swapped; this does not affect the beam trajectory in any way. In the considered case, when removing nine resonators (#8–#9, #12–#16 and #19–#20), we observe an ordinary waveguide. However, while preserving the structure, we can observe robust beam trajectory relative to the removal of resonator #3 from the TPhTI. In this case, resonators #8, #9, and #13–15 are used to compensate for the absence of resonator #3. Obviously, all nine internal resonators have the same purpose with respect to other restructurings. For example, the removal of resonator #22 is compensated for by resonators #12, #13, and #19. This is the main topological insulator feature; an ordinary waveguide lacks this feature and behaves unstably with respect to the mentioned restructurings.

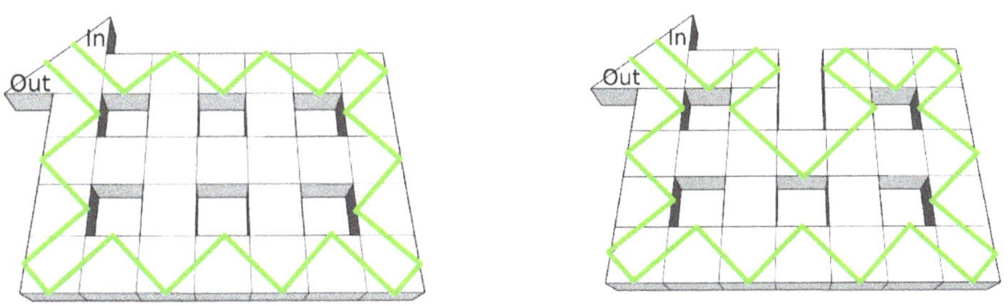

(a) (b)

Figure 2. (**a**) The beam trajectory inside a smooth TPhTI model. (**b**) The corrected trajectory inside the TPhTI without prism resonator #3, as denoted in (**b**). The model shows that the light beam successfully bends around the defect of TPhTI.

The topological properties of the prism array yield a robustness to restructuring. Similar photonic topological insulators are theoretically described in [17–19,24], and practically implemented [18,24] on several platforms. Due to the angle of incidence of the beam trajectory, the prism array is a practical implementation of Rudner's toy [17], which provides a simple example of a topological insulator. As a class of restructuring, we chose the addition and removal of resonators from the array. Deeper explanations behind the topology are included in the Supplementary Materials.

Figure 3 shows that the laser beam bypasses the defect and exits the triangular coupler. The beam trajectory passes along the surface of the TPhTI structure due to the effect of total internal reflection on the faces of the prism resonators and exits back, as evidenced by the bright spot in the upper left corner of Figure 3b. The propagating state persists against removing (Figure 3b,d) or attaching new prism resonators (Figure 3e).

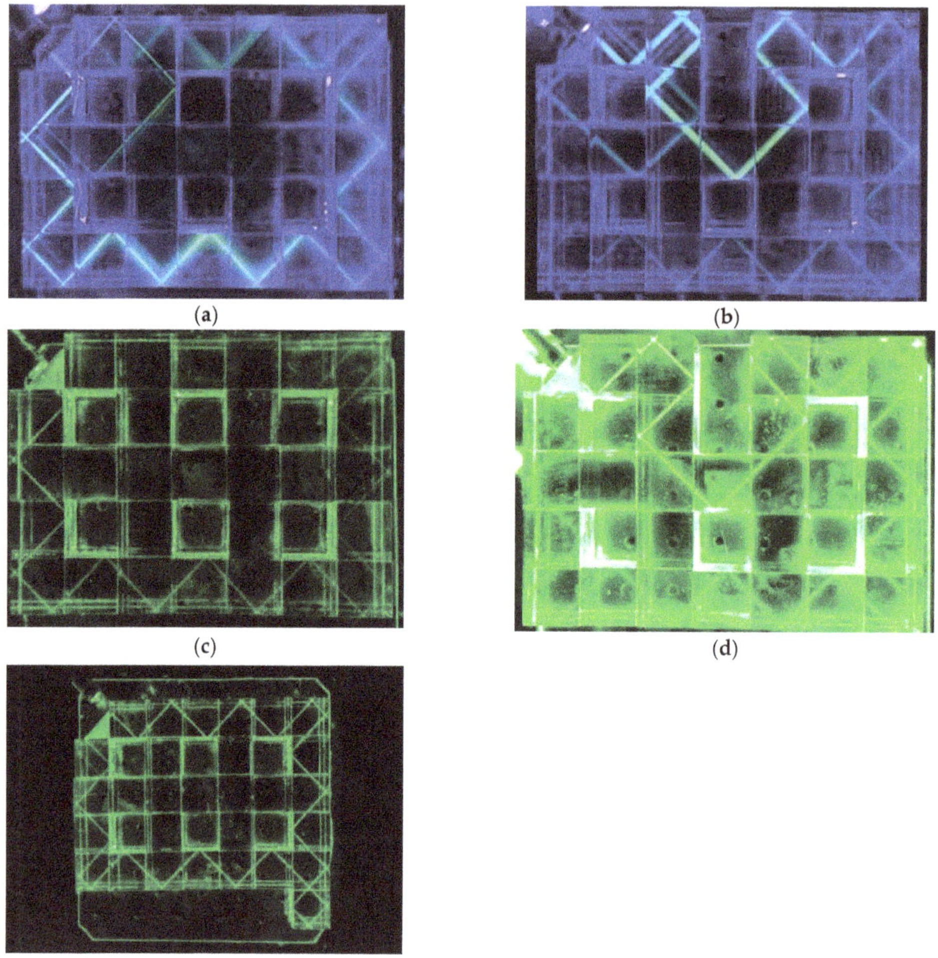

Figure 3. Beam trajectory photographs. The trajectory for smooth TPhTI (**a**,**c**) is robust against a defect obtained by removing prism resonator #3 (**b**,**d**) or attaching a new prism resonator below resonator #28 (**e**). Light beams for both a violet laser of 405 nm wavelength (**a**,**b**) and green laser of 532 nm (**c**–**e**) act in the same manner, proving the wavelength independence of the TPhTI principle.

Figure 4 shows that the presence of a topological state does not depend on prism resonator size. To demonstrate this scalability, the rectangular prism resonators were made in two versions. The first set was manufactured from quartz glass with a refractive index of 1.43, with linear dimensions 12 × 12 × 8 mm^3 (Figure 4a). Their linear dimensions were ensured by the Maksutov method [23] with 10 um accuracy. The second set was manufactured from Chinese crown glass K9 with a refractive index of 1.517 and linear dimensions 30 × 30 × 30 mm^3, with 30 um accuracy (Figure 4b).

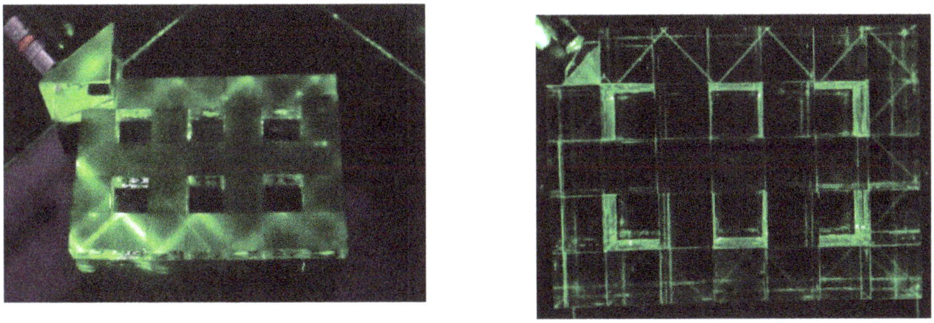

(a) (b)

Figure 4. Scalability of the TPhTI is manifested by comparing two resonator sizes: (**a**) $12 \times 12 \times 8$ mm^3. (**b**) $30 \times 30 \times 30$ mm^3. Both trajectories are homothetic and proportional to prism linear dimensions.

In Figure 5, the scalability of the TPhTI principle is stressed. First, it works for every wavelength in the visible range, as long as the refractive index is sufficiently high. Second, the resonator size was fixed at the centimeter scale; nonetheless, the principle is valid for arbitrary sizes larger than the wavelength and the beam cross-section. Please see additional trajectories for different wavelengths, resonator sizes, and tiling configurations in the Supplementary Materials.

(a) (b) (c)

Figure 5. Wavelength-independent beam trajectory. Laser beams of RGB-colors pass the same trajectory through a TPhTI made of seven prism resonators: (**a**) red 650 nm, (**b**) green 532 nm, and (**c**) blue 405 nm.

4. Discussion

In practical implementation, the TPhTI turned out to be sensitive to the following parameters of the experiment:

1. Propagation of the light beam at the angle of total internal reflection; the intensity of the beam weakens with small deviations of the angle of incidence of the beam through a triangular coupler.
2. The presence of dust particles, air bubbles in the immersion liquid layer between the prism resonators, or remnants of immersion liquid on the side faces of prism resonators scatters the light beam. Violation of the parallelism of the faces when connecting prism resonators and inaccuracies in the manufacturing of prism resonators deflect the direction of the light beam, as well as changing its aperture and cross-section shape. With a strong deviation, splitting of the light beam on the vertical edges of the prism resonators is possible.

Quantitatively, TPhTI performance is proportional to the beam intensity along the trajectory, as shown in Table 1 for the smooth TPhTI (28 resonators, Figure 3a,c) and defective TPhTI, with one resonator taken away (27 resonators, Figure 3b,d). The relative intensity is normalized to the intensity of the beam passed through the reference waveguide, consisting of eight resonators (Figure 5). Throughout the visible range, the smooth TPhTI secures more than half of the radiation, while the defective TPhTI loses 10–20% more. The best performance is seen in the middle wavelength. The first reason for this is the refractive index dispersion for the prism material, coupler material, and immersion liquid. The second reason is the difference in beam cross-section of the used sources. Altogether, these reasons are technical and have no deep physical impact.

Table 1. TPhTI performance is measured as the relative beam intensity of the edge trajectory.

Color	Red, 650 nm	Green, 532 nm	Blue, 405 nm
28 resonator TPhTI	51%	78%	68%
27 resonator TPhTI	43%	65%	48%

The effect of stability of the trajectory of the light beam relative to the defect introduced into the TPhTI is experimentally confirmed. Table 2 summarizes the stability of trajectories shown in Figures 3–5. Figure 4 shows the perfect scalability of the phenomenon. Therefore, the minus sign '-' declares that the linear dimensions of the prism resonator do not affect the trajectory stability. The refractive indices are not important until the total internal reflection and immersion conditions are satisfied. Therefore, they share the plus–minus sign '±'. The angle of beam incidence is exactly 45 degrees. This is critical, as the trajectory has to pass between certain corners or the beam will split and enter the array bulk. The partial impact of laser wavelength is shown in Table 1 and Figure 5.

Table 2. Experimental parameters and their impact on beam trajectory stability.

Name	Value	Value	Affects (+)/Does Not Affect (−) Stability
Linear dimensions of the prism resonator	$12 \times 12 \times 8$ mm^3	$30 \times 30 \times 30$ mm^3	- (see Figure 4)
Refractive index of the immersion liquid	1.45	1.45	±
Refractive index of quartz glass	1.43	1.517	±
Angle of total internal reflection	>44.37°	>41.24°	±
Angle of beam incidence	~45°	~45°	+
Laser wavelength	405/532/650 nm	405/532/650 nm	± (see Figure 5)

5. Conclusions

Here we focused on a new type of photonic topological insulator. We suggested a tiling photonic topological insulator constructed from identical prism resonators connected to each other. We experimentally demonstrated a topologically-protected propagating beam trajectory due to the disconnected faces of edge resonators. Then we removed or attached some prism resonators to destruct the trajectory. Nonetheless, the complete trajectory remained the same outside the destructed area. Overall, the main benefit is a novel platform for the practical implementation of Rudner's toy [17]: a tiling photonic topological insulator. Other benefits are as follows: experimental demonstration of the novel platform, its robustness, phase-free concept, and laser array application. As a direction for possible further research, it is worth noting that, along with quadrangular TPhTIs, triangular and hexagonal prism resonator tiling can be experimentally implemented by zero gap, in accordance with the theoretical prediction [19]. One of the advantages of the hexagonal TPhTI is that the reflection angle is 60°. Therefore, this type of insulator can be experimentally implemented with optical material with a refractive index less than 1.43; hence, no less than 1.15. On the other hand, the simplest geometric considerations show that, under certain conditions in a hexagonal TPhTI, a diagonal transition occurs in fewer jumps between resonators than in a quadrangular TPhTI. The TPhTI can be described in the language of trajectories, as well as in the language of differential wave equations natural for graphene-like materials [25], secondary quantization operators, or cellular automata [26]. For some certain incidence angles, the edge solutions reversed the direction upon scattering by edge defects. This case may be indicative of the existence of higher-order states [27].

This tiling is suggested for active topological photonic devices and laser arrays, while topological features force unidirectional propagation. In the visible range, the magneto-optic effects are generally weak, which prohibits the conventional unidirectional propagation mechanism. Instead, the imbedding of gain and loss materials inside the resonator provides another type of unidirectionality [1,2]. The conducted research paves the way to transfer the gain–loss unidirectional mechanism from photonic crystals and waveguides to phase-free theory for macroscopic prism resonators.

Supplementary Materials: The following supporting information can be downloaded at: https://www.mdpi.com/article/10.3390/app13064004/s1, Figures S1–S7: beam trajectory when passing through the TPhTI; Figures S4 and S5: trajectory that bends around the defect in the structure of the TPhTI; Figures S6 and S7: trajectory that bends around the defect in the structure of the TPhTI with resonators of a different size; Figures S8–S10: operation of Snell's law; Figure S11: scattering matrix formalism to explain topological protection [17,24,28–31].

Author Contributions: Sample preparation, experimental setup, and measurement, P.N.K., N.V.R. and I.V.T.; software, D.P.F. and I.V.T.; validation, P.N.K.; conceptualization and methodology, D.P.F. and I.V.T.; writing—review and editing, P.N.K., D.P.F., N.V.R. and I.V.T.; supervision, I.V.T. All authors have read and agreed to the published version of the manuscript.

Funding: This research was funded by the Russian Science Foundation (project no. 22-42-08003).

Data Availability Statement: The data presented in this study are available upon reasonable request from the corresponding author.

Acknowledgments: The authors are grateful to V.M. Sventitskii for useful discussions and materials used for experiment.

Conflicts of Interest: The authors declare no conflict of interest.

References

1. Bandres, M.A.; Wittek, S.; Harari, G.; Parto, M.; Ren, J.; Segev, M.; Christodoulides, D.N.; Khajavikhan, M. Topological insulator laser: Experiments. *Science* **2018**, *359*, eaar4005. [CrossRef] [PubMed]
2. Harari, G.; Bandres, M.A.; Lumer, Y.; Rechtsman, M.C.; Chong, Y.D.; Khajavikhan, M.; Christodoulides, D.N.; Segev, M. Topological insulator laser: Theory. *Science* **2018**, *359*, eaar4003. [CrossRef] [PubMed]
3. Zeng, Y.; Chattopadhyay, U.; Zhu, B.; Qiang, B.; Li, J.; Jin, Y.; Li, L.; Davies, A.G.; Linfield, E.H.; Zhang, B.; et al. Electrically pumped topological laser with valley edge modes. *Nature* **2020**, *578*, 246–250. [CrossRef] [PubMed]
4. Dikopoltsev, A.; Harder, T.H.; Lustig, E.; Egorov, O.A.; Beierlein, J.; Wolf, A.; Lumer, Y.; Emmerling, M.; Schneider, C.; Höfling, S.; et al. Topological insulator vertical-cavity laser array. *Science* **2021**, *373*, 1514–1517. [CrossRef] [PubMed]
5. Shao, Z.-K.; Chen, H.-Z.; Wang, S.; Mao, X.-R.; Yang, Z.-Q.; Wang, S.-L.; Wang, X.-X.; Hu, X.; Ma, R.-M. A high-performance topological bulk laser based on band-inversion-induced reflection. *Nat. Nanotechnol.* **2020**, *15*, 67–72. [CrossRef]
6. Yang, Z.-Q.; Shao, Z.-K.; Chen, H.-Z.; Mao, X.-R.; Ma, R.-M. Spin-Momentum-Locked Edge Mode for Topological Vortex Lasing. *Phys. Rev. Lett.* **2020**, *125*, 013903. [CrossRef]
7. Schomerus, H. Topologically protected midgap states in complex photonic lattices. *Opt. Lett.* **2013**, *38*, 1912. [CrossRef]
8. St-Jean, P.; Goblot, V.; Galopin, E.; Lemaître, A.; Ozawa, T.; Le Gratiet, L.; Sagnes, I.; Bloch, J.; Amo, A. Lasing in topological edge states of a one-dimensional lattice. *Nat. Photonics* **2017**, *11*, 651–656. [CrossRef]
9. Parto, M.; Wittek, S.; Hodaei, H.; Harari, G.; Bandres, M.A.; Ren, J.; Rechtsman, M.C.; Segev, M.; Christodoulides, D.N.; Khajavikhan, M. Edge-Mode Lasing in 1D Topological Active Arrays. *Phys. Rev. Lett.* **2018**, *120*, 113901. [CrossRef]
10. Ishida, N.; Ota, Y.; Lin, W.; Byrnes, T.; Arakawa, Y.; Iwamoto, S. A large-scale single-mode array laser based on a topological edge mode. *Nanophotonics* **2022**, *11*, 2169–2181. [CrossRef]

11. Hasan, M.Z.; Kane, C.L. Colloquium: Topological insulators. *Rev. Mod. Phys.* **2010**, *82*, 3045–3067. [CrossRef]
12. Thouless, D.J.; Kohmoto, M.; Nightingale, M.P.; den Nijs, M. Quantized Hall conductance in a two-dimensional periodic potential. *Phys. Rev. Lett.* **1982**, *49*, 405. [CrossRef]
13. Cheng, X.; Jouvaud, C.; Ni, X.; Mousavi, S.H.; Genack, A.Z.; Khanikaev, A.B. Robust reconfigurable electromagnetic pathways within a photonic topological insulator. *Nat. Mater.* **2016**, *15*, 542–548. [CrossRef]
14. Haldane, F.D.M.; Raghu, S. Possible Realization of Directional Optical Waveguides in Photonic Crystals with Broken Time-Reversal Symmetry. *Phys. Rev. Lett.* **2008**, *100*, 013904. [CrossRef]
15. Mong, R.S.K.; Shivamoggi, V. Edge states and the bulk-boundary correspondence in Dirac Hamiltonians. *Phys. Rev. B* **2011**, *83*, 125109. [CrossRef]
16. Khanikaev, A.B.; Shvets, G. Two-dimensional topological photonics. *Nat. Photonics* **2017**, *11*, 763–773. [CrossRef]
17. Rudner, M.S.; Lindner, N.H.; Berg, E.; Levin, M. Anomalous Edge States and the Bulk-Edge Correspondence for Periodically Driven Two-Dimensional Systems. *Phys. Rev. X* **2013**, *3*, 031005. [CrossRef]
18. Leykam, D.; Yuan, L. Topological phases in ring resonators: Recent progress and future prospects. *Nanophotonics* **2020**, *9*, 4473–4487. [CrossRef]
19. Fedchenko, D.P.; Kim, P.N.; Timofeev, I.V. Photonic Topological Insulator Based on Frustrated Total Internal Reflection in Array of Coupled Prism Resonators. *Symmetry* **2022**, *14*, 2673. [CrossRef]
20. Yang, Z.; Lustig, E.; Harari, G.; Plotnik, Y.; Lumer, Y.; Bandres, M.A.; Segev, M. Mode-Locked Topological Insulator Laser Utilizing Synthetic Dimensions. *Phys. Rev. X* **2020**, *10*, 011059. [CrossRef]
21. Liu, Y.G.N.; Wei, Y.; Hemmatyar, O.; Pyrialakos, G.G.; Jung, P.S.; Christodoulides, D.N.; Khajavikhan, M. Complex skin modes in non-Hermitian coupled laser arrays. *Light Sci. Appl.* **2022**, *11*, 336. [CrossRef] [PubMed]
22. Liu, Y.G.N.; Jung, P.S.; Parto, M.; Christodoulides, D.N.; Khajavikhan, M. Gain-induced topological response via tailored long-range interactions. *Nat. Phys.* **2021**, *17*, 704–709. [CrossRef]
23. Maksutov, D.D. New Catadioptric Meniscus Systems. *J. Opt. Soc. Am.* **1944**, *34*, 270. [CrossRef]
24. Gao, F.; Gao, Z.; Shi, X.; Yang, Z.; Lin, X.; Xu, H.; Joannopoulos, J.D.; Soljačić, M.; Chen, H.; Lu, L.; et al. Probing topological protection using a designer surface plasmon structure. *Nat. Commun.* **2016**, *7*, 11619. [CrossRef] [PubMed]
25. Grushevskaya, H.V.; Krylov, G.G.; Kruchinin, S.P.; Vlahovic, B.; Bellucci, S. Electronic properties and quasi-zero-energy states of graphene quantum dots. *Phys. Rev. B* **2021**, *103*, 235102. [CrossRef]
26. Farrelly, T. A review of quantum cellular automata. *Quantum* **2020**, *4*, 368. [CrossRef]
27. Li, M.; Zhirihin, D.; Gorlach, M.; Ni, X.; Filonov, D.; Slobozhanyuk, A.; Alù, A.; Khanikaev, A.B. Higher-order topological states in photonic kagome crystals with long-range interactions. *Nat. Photonics* **2020**, *14*, 89–94. [CrossRef]
28. Pasek, M.; Chong, Y.D. Network models of photonic Floquet topological insulators. *Phys. Rev. B—Condens. Matter Mater. Phys.* **2014**, *89*, 075113. [CrossRef]
29. Shalaev, M.I.; Walasik, W.; Tsukernik, A.; Xu, Y.; Litchinitser, N.M. Robust topologically protected transport in photonic crystals at telecommunication wavelengths. *Nat. Nanotechnol.* **2019**, *14*, 31–34. [CrossRef]
30. Joannopoulos, J.; Johnson, S.; Winn, J.; Meade, R.D. *Photonic Crystals: Molding the Flow of Light*, 2nd ed.; Princeton University: Princeton, NJ, USA, 2008; p. 305.
31. Vetrov, S.Y.; Timofeev, I.V.; Shabanov, V.F. Localized modes in chiral photonic structures. *Physics-Uspekhi* **2020**, *63*, 33. [CrossRef]

applied sciences

MDPI

Article

Full Explicit Numerical Modeling in Time-Domain for Nonlinear Electromagnetics Simulations in Ultrafast Laser Nanostructuring

Enrique Moreno *, Huu Dat Nguyen, Razvan Stoian and Jean-Philippe Colombier *

Laboratoire Hubert Curien UMR5516, Institute of Optics Graduate School, CNRS, UJM-Saint-Etienne, University of Lyon, F-42023 St-Etienne, France; huu.dat.nguyen@univ-st-etienne.fr (H.D.N.); razvan.stoian@univ-st-etienne.fr (R.S.)

* Correspondence: enrique@moreno.ws (E.M.); jean.philippe.colombier@univ-st-etienne.fr (J.-P.C.)

Abstract: The purpose of this paper is to present a new and accurate, fully explicit finite-difference time-domain method for modeling nonlinear electromagnetics. The approach relies on a stable algorithm based on a general vector auxiliary differential equation in order to solve the curl Maxwell's equation in a frequency-dependent and nonlinear medium. The energy conservation and stability of the presented scheme are theoretically proved. The algorithms presented here can accurately describe laser pulse interaction with metals and nonlinear dielectric media interfaces where Kerr and Raman effects, as well as multiphoton ionization and metal dispersion, occur simultaneously. The approach is finally illustrated by simulating the nonlinear propagation of an ultrafast laser pulse through a dielectric medium transiently turning to inhomogeneous metal-like states by local free-electron plasma formation. This free carrier generation can also be localized in the dielectric region surrounding nanovoids and embedded metallic nanoparticles, and may trigger collective effects depending on the distance between them. The proposed numerical approach can also be applied to deal with full-wave electromagnetic simulations of optical guided systems where nonlinear effects play an important role and cannot be neglected.

Keywords: finite-difference time-domain method (FDTD); nonlinear propagation; Raman effect simulation; Kerr effect simulation; light propagation in a photoionizable media; plasma; Maxwell equations solver; laser pulse interaction; general vector auxiliary differential equation (GVADE)

Citation: Moreno, E.; Nguyen, H.D.; Stoian, R.; Colombier, J.-P. Full Explicit Numerical Modeling in Time-Domain for Nonlinear Electromagnetics Simulations in Ultrafast Laser Nanostructuring. *Appl. Sci.* **2021**, *11*, 7429. https://doi.org/10.3390/app11167429

Academic Editor: Gennady M. Mikheev

Received: 9 July 2021
Accepted: 9 August 2021
Published: 12 August 2021

1. Introduction

The finite difference time-domain (FDTD) method enables the computation and modeling of light propagation and scattering processes in linear and nonlinear dispersive materials possessing wavelength-dependent and intensity-dependent properties. This ranks the method among the most universal and powerful numerical tools for optics, electrodynamics, antennas, and waveguides theory. In this context, the general vector auxiliary differential equation (GVADE) method, which has been reported in [1] and improved in [2], is one of the most popular methods to deal with nonlinear media. However, the method suffers from some drawbacks that limit its speed and stability due to the semi-implicit updating relation for the current polarization densities. In summary, the method is not appropriate or recommended for a broad range of problems, especially those that address the designs with complex 3D topology, where iterative solutions of coupled nonlinear equations can be unstable and resource-demanding [3]. Some direct or explicit methods take into consideration the Kerr effect in the Born–Oppenheimer approximation [4]. The Kerr effect arises from the nonlinear third-order electric susceptibility. In [3], the authors take into consideration the Kerr effect in an explicit form. Other direct methods [5] require excessive computational burdens, especially for the consideration of complex geometries. When metal–dielectric interfaces are treated, especially at nanometric scales, plasmonic effects appear [6,7] and generate high concentrations or high local densities of the electromagnetic field [8]. Since dielectrics experience nonlinear optical processes, the treatment of these regions in the interfaces is particularly challenging.

In this manuscript, we develop an algorithm that permits solving a nonlinear dielectric medium in which Kerr and Raman's effects [9] occur simultaneously. The Raman effect arises from the scattering caused by the conversion of photon energies into vibrational energy of molecules. The scattered photons have less energy than the incident photons corresponding, therefore lowering frequency fields [2]. We also assume that this takes place in an inhomogeneous dielectric environment that can potentially include metallic heterogeneities with specific optical properties. Along with the laser propagation, provided that the intensity is sufficient to ensure nonlinear absorption, the medium can also turn into a metallic state as a free electron population is generated during multiphoton absorption (MPA) or tunneling and avalanche ionization processes. It is well known that the electron transition from the valence band (VB) to the conduction band (CB) is followed by various processes, including heating in the CB, impact ionization, and various kinds of electron collisions. The purpose of this work consists of computing all these events in a full explicit form. The carrier generation rate $G(\vec{r},t)$ has been computed owing to the BVkP approach implying Bloch–Volkov states [10]. The algorithms presented in this manuscript are especially suitable for simulations dealing with confinements and guiding high-energy pulsed beams. In summary, in this work, we coupled Maxwell equations and the electric charge carriers conservation equation in which the quantum ionization and recombination are governed by BVkP self-consistently.

In the next sections, by following and enriching some pioneers' works [11–13], we detail a fully explicit, stable algorithm based on GVADE FDTD, able to solve the curl Maxwell's equation in a frequency-dependent way for a nonlinear medium. In order to illustrate the combination of all these nonlinear effects occurring simultaneously in a simulation, we have modeled a Bessel beam traveling through dielectric fused silica, where some metal inclusions are modeled in a similar form as in Ref. [14]. In Section 2, we introduce the theoretical model, which describes the nonlinear effects. Section 3 presents the algorithms that discretize in time and space the main magnitudes, electromagnetic fields and electric charge. Appendix A is devoted to the stability condition assessment for these algorithms, which imposed numerical limits to the simulation performance. These limits are studied in Section 4, where we specify the validity framework considering natural limits imposed by physics. In Section 5 we present the simulation outcomes provided by the full numerical modelling and in Section 6 the conclusions are drawn. Finally, the stability conditions are derived in Appendix A and the used computational resources are presented in Appendix B.

2. Modeling Electromagnetic Fields in Nonlinear Media

The Equations (1)–(3) model the laser beam propagation by means of the electromagnetic fields $\vec{E}(\vec{r},t)$ and $\vec{H}(\vec{r},t)$. Throughout the light propagation path, fields generate and couple with the density of free charges as free electrons $n_{fe^-}(\vec{r},t)$, which in turn interact with the local field. In the given form, nonlinearities are determined by the electric displacement field $\vec{D}(\vec{r},t)$ in the Ampère-Maxwell Equation (1). Meanwhile, the evolution of the free-electron density or medium conductivity is described by Equation (3), where $G(\vec{r},t)$ is the carrier generation rate and $R(\vec{r},t)$ is the carrier recombination rate.

$$\nabla \wedge \vec{H}(\vec{r},t) = \frac{\partial \vec{D}(\vec{r},t)}{\partial t} \tag{1}$$

$$\nabla \wedge \vec{E}(\vec{r},t) = -\mu \frac{\partial \vec{H}(\vec{r},t)}{\partial t} \tag{2}$$

$$\frac{\partial n_{fe^-}(\vec{r},t)}{\partial t} = G(\vec{r},t) - R(\vec{r},t) \tag{3}$$

The electric displacement field is composed by the electric field contribution plus the polarization vectors $\vec{D}(\vec{r},t) = \varepsilon_0\varepsilon_\infty\vec{E}(\vec{r},t) + \sum\vec{P}(\vec{r},t)$ [2]. The magnitude ε_∞ is the non-unity high-frequency relative permittivity of the media. On the right hand side of the displacement electric field, the sum of polarization vectors in the model that we are considering $\sum\vec{P}(\vec{r},t)$ is the contribution of four terms, the metal polarization $\vec{P}_{Metal}(\vec{r},t)$, the Kerr effect polarization $\vec{P}_{Kerr}(\vec{r},t)$, the Raman polarization $\vec{P}_{Raman}(\vec{r},t)$ and the free charge polarization vector $\vec{P}_{fe^-}(\vec{r},t)$ such as $\sum\vec{P}(\vec{r},t) = \vec{P}_{Metal}(\vec{r},t) + \vec{P}_{Kerr}(\vec{r},t) + \vec{P}_{Raman}(\vec{r},t) + \vec{P}_{fe^-}(\vec{r},t)$ with the detail of each term below:

$$\vec{P}_{Metal}(\vec{r},t) = \frac{\epsilon_0 a}{j\omega + b}\vec{E}(\vec{r},t) \qquad a,b \in \mathbb{C} \tag{4}$$

$$\vec{P}_{Kerr}(\vec{r},t) = \alpha\epsilon_0\chi_0^{(3)}\vec{E}(\vec{r},t)\int_{-\infty}^{t}\delta(t-\tau)|\vec{E}(\vec{r},\tau)|^2 d\tau \tag{5}$$

$$\vec{P}_{Raman}(\vec{r},t) = \Psi\vec{E}(\vec{r},t)\int_{-\infty}^{t} e^{\frac{-(t-\tau)}{\tau_2}}\sin(\frac{t-\tau}{\tau_1})U(t-\tau)|\vec{E}(\vec{r},\tau)|^2 d\tau \tag{6}$$

$$\vec{P}_{fe^-}(\vec{r},t) = \int_{-\infty}^{t}\sigma_{fe^-}(\vec{r},t)\vec{E}(\vec{r},t)d\tau \tag{7}$$

where $\Psi = \varepsilon_0\chi^{(3)}(1-\alpha)\frac{\tau_1^2+\tau_2^2}{\tau_1\tau_2^b yte,intent(in)::types2}$ is a coefficient gathering parameters depending on the considered medium. Expressions Equations (5) and (6) providing the Kerr effect polarization and Raman polarization, respectively, are taken from [9]. Here, α is a real-valued constant in the range $\alpha \in [0,1]$ that parameterizes the relative strengths of Kerr and Raman contributions, $\delta(t)$ refers to the Dirac delta function that models the Kerr nonresonant virtual electronic transitions, which is medium dependent and in the order of few femtoseconds for fused silica, $U(t)$ is the Heaviside unit step function, $e^{\frac{-t}{\tau_2}}\sin(\frac{t}{\tau_1})$ models the impulse response of a single Lorentzian relaxation centered on the optical phonon frequency $1/\tau_1$ and having a bandwidth of $1/\tau_2$. Finally $\chi_0^{(3)}$ is the strength of the third-order nonlinear electric susceptibility.

The upper limit of the integrations in Equations (5)–(7) extends only up to t because the response function must be zero for $\tau > t$ to ensure causality [3]. Moreover, free-charge polarization is introduced by using a free-charge density of current, which will be defined later. Finally, to model the metal, we consider a complex electrical permittivity $\tilde{\varepsilon}(\omega) = \frac{a}{j\omega+b}$ where a and b are complex numbers that fit experimental data [15]. To model the dispersive media, we assume an electric permittivity, which is frequency-dependent as the calculations are devoted to ultrafast laser propagation, with bandwidth-limited pulses that exhibit wavelength dispersion.

In ref. [2], the general vector auxiliary differential equation technique allows considering nonlinearities or dispersive characteristics of a medium. They are introduced into the Ampère–Maxwell equation by means of current densities. These currents correspond intrinsically to the temporal variation of the polarization vectors. Thereby, the following Equations (8)–(11) can be derived to correlate the currents $\vec{J}_{Metal}(\vec{r},t) = \frac{\partial\vec{P}_{Metal}(\vec{r},t)}{\partial t}$, $\vec{J}_{Kerr}(\vec{r},t) = \frac{\partial\vec{P}_{Kerr}(\vec{r},t)}{\partial t}$, $\vec{J}_{Raman}(\vec{r},t) = \frac{\partial\vec{P}_{Raman}(\vec{r},t)}{\partial t}$, $\vec{J}_{fe^-}(\vec{r},t) = \frac{\partial\vec{P}_{fe^-}(\vec{r},t)}{\partial t}$ with the electric field:

$$\vec{J}_{Metal}(\vec{r},t) = \frac{a\epsilon_0}{b}\frac{\vec{E}(\vec{r},t)}{\partial t} - \frac{1}{b}\frac{\vec{J}_{Metal}(\vec{r},t)}{\partial t} \tag{8}$$

$$\vec{J}_{Kerr}(\vec{r},t) = \alpha\epsilon_0\chi_0^{(3)}\frac{\partial}{\partial t}\left[\vec{E}(\vec{r},t)\int_0^t\delta(t-\tau)|\vec{E}(\vec{r},\tau)|^2 d\tau\right] \tag{9}$$

$$\vec{J}_{Raman}(\vec{r},t) = \Psi\frac{\partial}{\partial t}\left[\vec{E}(\vec{r},t)\int_0^t e^{\frac{-(t-\tau)}{\tau_2}}\sin(\frac{t-\tau}{\tau_1})U(t-\tau)|\vec{E}(\vec{r},\tau)|^2 d\tau\right] \tag{10}$$

$$\vec{J}_{fe^-}(\vec{r},t) = \sigma_{fe^-}(\vec{r},t)\vec{E}(\vec{r},t) \tag{11}$$

We define a free-charge density of current as an Ohmic current, where the conductivity $\sigma_{fe^-}(\vec{r},t) = q\mu_{fe^-} n_{fe^-}(\vec{r},t)$ is given by the following frequency dependent expression:

$$\sigma(\vec{r},\omega) = \frac{q\mu_{fe^-} f_d(|\vec{E}|) n_{ph-e^-}(|\vec{E}|)}{j\omega + f_r(|\vec{E}|) - f_a(|\vec{E}|) - f_\eta(|\vec{E}|)} \tag{12}$$

where q is the electron charge, μ_{fe^-} the free carriers of charge mobility, $n_{fe^-}(\vec{r},t)$ the free-charge density and $n_{ph-e^-}(\vec{r},t)$ the photo-ionized electron density. This definition for the density of current Equation (11) is similar to Equation (9) in [10]. In Equation (13), we employ free-charge which is governed by Equation (3), or generation Equation (13) and recombination Equation (14) rates. In [10], the authors explain that electron density $n_{ph-e^-}(\vec{r},t)$ produced in the conduction band only describes the interaction of an electron with the laser electric field and accounts for the lattice periodicity, but does not account for possible collisions with phonons, ions, or other electrons. Due to these collisions, the coherence between the excited electron to the CB and its parent ion is destroyed and may introduce deviations from the above-predicted evolution of the electron density in the CB with respect to time. Indeed, without collisions, the produced electron density was shown to oscillate with time as a conduction electron may go back and forth to the VB through the action of the electromagnetic field [10,16]. Obviously, in an oscillation, the energy stays constant [17]. Hence, the authors of [10] defined a free state as a state where the electrons can be heated by inverse bremsstrahlung (IB) and are decorrelated from the parent ion. The density of charge associated with this free state is called here density of free-charge $n_{fe^-}(\vec{r},t)$, and its evolution along the time is modeled by Equation (3). Nevertheless, the μ_{fe^-}, the free carriers of charge mobility, then we employ experimental data, which allow the determination of this parameter. The method is explained in detail in Section 5.

$$G(\vec{r},t) = f_d(\vec{r},t) n_{ph-e^-}(\vec{r},t) + n_{fe^-}(\vec{r},t)\left[f_a(\vec{r},t) + f_\eta(\vec{r},t)\right] \tag{13}$$

$$R(\vec{r},t) = f_r(\vec{r},t) n_{fe^-}(\vec{r},t) \tag{14}$$

In the carrier generation rate, the term associated with the photo-ionized electron density $n_{ph-e^-}(\vec{r},t)$ produced in the conduction band is a function of $|\vec{E}(\vec{r},t)|$. From Equation (5) in [10] we know that $\frac{\partial}{\partial t}\left[n_{ph-e^-}(\vec{r},t) - N_0 \sum_c |T_{cv}(t)|^2\right] = 0$ and hence $n_{ph-e^-}(\vec{r},t) - N_0 \sum_c |T_{cv}(t)|^2 = cte$. If this constant $cte = 0$, then $n_{ph-e^-}(\vec{r},t) = 0$ when $\vec{E}(\vec{r},t) = \bar{0}$. Combining this result with Equations (3) and (4) in [10], we calculate $n_{ph-e^-}(\vec{r},t)$ by Equation (15).

$$n_{ph-e^-}(\vec{r},t) = \sum_{\xi_i=0}^{W_f} \left| \frac{-\sqrt{E_g}\int_0^t \frac{|\vec{E}(\vec{r},\vartheta)| e^{-j\vartheta(E_g+\xi_i)}}{\left(1-\frac{j}{\hbar}\int_0^\vartheta |\vec{E}(\vec{r},\varrho)| d\varrho\right)^2} d\vartheta}{N_0^{-1}(E_g+\xi_i)\sqrt{2m_{vc}^*}} \right|^2 \tag{15}$$

Finally the functions related to the generation by ionization $f_d(\vec{r},t)$, impact $f_\eta(\vec{r},t)$ and avalanche $f_a(\vec{r},t)$, as well as the function related to the recombination of the free-charge $f_r(\vec{r},t)$, are given by the Equations (16)–(19), respectively.

$$f_d(\vec{r},t) = \begin{cases} \frac{1}{\tau_d} & \text{if } |\vec{E}(\vec{r},t)| \geq \frac{E_{max}}{20} \\ \frac{20|\vec{E}(\vec{r},t)|}{\tau_d E_{max}} & \text{if } |\vec{E}(\vec{r},t)| < \frac{E_{max}}{20} \end{cases} \tag{16}$$

$$f_\eta(\vec{r},t) = \begin{cases} \eta & \text{if } |\vec{E}(\vec{r},t)| \geq \frac{E_{max}}{20} \\ \frac{20|\vec{E}(\vec{r},t)|\eta}{E_{max}} & \text{if } |\vec{E}(\vec{r},t)| < \frac{E_{max}}{20} \end{cases} \tag{17}$$

$$f_a(\vec{r},t) = \begin{cases} \frac{\Theta|\vec{E}(\vec{r},t)|^2}{\zeta} & \text{Harmonic field} \\ \Theta\left(\vec{E}(\vec{r},t) \wedge \vec{H}(\vec{r},t)\right) \cdot \hat{z} & \text{No harmonic field} \end{cases} \quad (18)$$

$$f_r(\vec{r},t) = \begin{cases} \frac{1}{\tau_r} & \text{if } |\vec{E}(\vec{r},t)| \geq \frac{E_{max}}{20} \\ \frac{20|\vec{E}(\vec{r},t)|}{\tau_r E_{max}} & \text{if } |\vec{E}(\vec{r},t)| < \frac{E_{max}}{20} \end{cases} \quad (19)$$

In Equations (16)–(19), we find that the factor of maximum electric field reduction, which is suggested twenty times by [10], is taken into consideration. Nevertheless, when the photoionization process is considered in a truncated region where the magnitude of $|\vec{E}(\vec{r},t)|$ is from the beginning higher than the threshold $\frac{E_{max}}{20}$ reference value, these logical conditions can be suppressed in the nested loops, and this increases the algorithm performance. In other words, this simplification is possible when we assume the description with nonlinear effects in a particular region of the computational domain. So, the electromagnetic field that arrives at that region has a magnitude higher than the threshold $\frac{E_{max}}{20}$.

In Equation (18), we assume that $\frac{|\vec{E}(\vec{r},t)|^2}{\zeta}$ is the fluence with ζ being the medium impedance. However, this is a particular case only valid when the electromagnetic field is a harmonic function in time. In general, and assuming a propagation along the z-axis, the power per surface is given by the Poynting vector.

3. Algorithms

The polarization vector due to the Kerr effect at the time $t = \Delta_t\left(s - \frac{1}{2}\right)$ can be written as $\vec{P}^{s-\frac{1}{2}}_{Kerr} = \alpha\epsilon_0\chi_0^{(3)}|\vec{E}^{s-\frac{1}{2}}|^2\vec{E}^{s-\frac{1}{2}}$; however, afterwards, at the instant $\Delta_t\left(s + \frac{1}{2}\right)$, in order to define a full explicit approach, we have to make the approximation $\int_0^{\left(s+\frac{1}{2}\right)\Delta_t} \delta(t - \tau)|\vec{E}(\vec{r},\tau)|d\tau \simeq \int_0^{\left(s-\frac{1}{2}\right)\Delta_t} \delta(t-\tau)|\vec{E}(\vec{r},\tau)|d\tau$. This approximation was demonstrated in [3] and leads to:

$$\vec{P}^{s+\frac{1}{2}}_{Kerr} \simeq \alpha\epsilon_0\chi_0^{(3)}\vec{E}^{s+\frac{1}{2}} \int_0^{\left(s-\frac{1}{2}\right)\Delta_t} \delta(t-\tau)|\vec{E}(\vec{r},\tau)|d\tau$$

This approximation $\vec{P}^{s+\frac{1}{2}}_{Kerr} \simeq \alpha\epsilon_0\chi_0^{(3)}|\vec{E}^{s-\frac{1}{2}}|^2\vec{E}^{s+\frac{1}{2}}$ is valid under particular physical conditions explained in Section 4. To compute the nonlinear Raman effect, we make an analogous approximation. Considering that in this case there is no instantaneous cause–effect relationship, it is evident that the approximation is justified [18]. Therefore, by defining the scalar magnitude $|\vec{Q}_a(\vec{r},t)|$ that is defined from the vector $\vec{Q}_a(\vec{r},t)$, which could be considered as an instantaneous anisotropic Raman conductivity, we can discretize the Raman polarization vectors at the instants $\Delta_t\left(s - \frac{1}{2}\right)$ and $\Delta_t\left(s + \frac{1}{2}\right)$ as follows: $\vec{P}^{s-\frac{1}{2}}_{Raman} = |\vec{Q}_a^{s-\frac{1}{2}}|\vec{E}^{s-\frac{1}{2}}$ and $\vec{P}^{s+\frac{1}{2}}_{Raman} = |\vec{Q}_a^{s-\frac{1}{2}}|\vec{E}^{s+\frac{1}{2}}$. Where $|\vec{Q}_a^{s+\frac{1}{2}}|$ is updated from $|\vec{Q}_a^{s-\frac{1}{2}}|$ by means of:

$$|\vec{Q}_a^{s+\frac{1}{2}}| = |\vec{Q}_a^{s-\frac{1}{2}}| + \bar{\bar{g}}((s - nt(k))\Delta_t)|\vec{E}^{s-\frac{1}{2}}|^2$$

where the Raman response function for a particular medium is modeled by $\bar{\bar{g}}(t) = \Psi e^{-\frac{t}{\tau_2}}\sin\left(\frac{t}{\tau_1}\right)u(t)$ and the time delay $nt(k) = nt_0 - \frac{H_c(k-k_0)\Delta_z n_{medium}}{c_0\Delta t}$, being $nt_0 = \frac{H_c}{c_0\Delta_t}\sum_{i=1}^{N}\ell_i n_i$, depends on the refractive index of the media n_i as well as the location of the electromagnetic field in the media (Figure 1). Here $u(t)$ is the Heaviside step function:

$$u(t) = \begin{cases} 0 & \text{if } t \leq 0 \\ 1 & \text{if } t > 0 \end{cases}$$

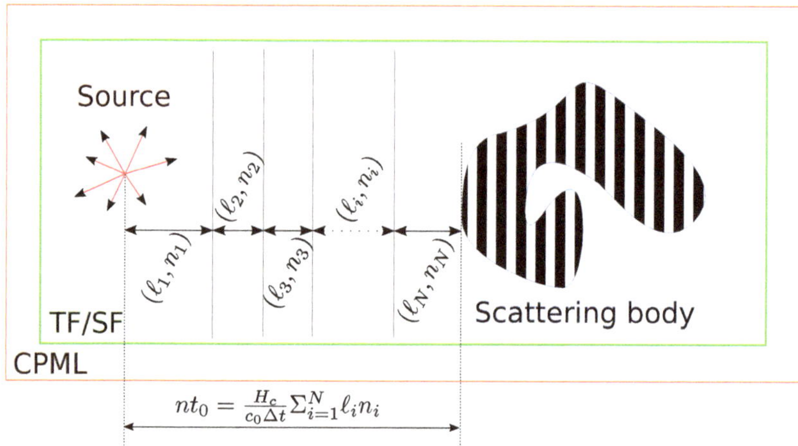

Figure 1. Scheme of the calculation box, illustrating the delay $nt_0(k)$ that depends on the refractive index of the media n_i as well as the location of the electromagnetic field in the media.

In this way, we obtain the Raman current density and the Kerr current density:

$$\vec{J}^n_{Kerr} = \frac{\vec{P}^{s+\frac{1}{2}}_{Kerr} - \vec{P}^{s-\frac{1}{2}}_{Kerr}}{\Delta_t} = \frac{\alpha\epsilon_0\chi_0^{(3)}|\vec{E}^{s-\frac{1}{2}}|^2}{\Delta_t}\left(\vec{E}^{s+\frac{1}{2}} - \vec{E}^{s-\frac{1}{2}}\right)$$

$$\vec{J}^n_{Raman} = \frac{\vec{P}^{s+\frac{1}{2}}_{Raman} - \vec{P}^{s-\frac{1}{2}}_{Raman}}{\Delta_t} = \frac{|\vec{Q}_a^{s-\frac{1}{2}}|}{\Delta_t}\left(\vec{E}^{s+\frac{1}{2}} - \vec{E}^{s-\frac{1}{2}}\right)$$

The current density updating associated to the metal is given by the equation:

$$\vec{J}^{s+\frac{1}{2}}_{Metal} = g_a \vec{J}^{s-\frac{1}{2}}_{Metal} + g_b\left(\vec{E}^{s+\frac{1}{2}} - \vec{E}^{s-\frac{1}{2}}\right)$$

where $g_a = \mathcal{R}\left[\frac{2-b\Delta_t}{2+b\Delta_t}\right]$ and $g_b = \mathcal{R}\left[\frac{a\epsilon_0}{2+b\Delta_t}\right]$. The operator $\mathcal{R}[]$ calculates the real part of these complex numbers. Finally, the electric field updating is given by the algorithm:

$$\vec{E}^{s+\frac{1}{2}} = \vec{E}^{s-\frac{1}{2}} + C_B^s\left(\nabla \wedge \vec{H}^s - g_c \vec{J}^{s-\frac{1}{2}}_{Metal}\right)$$

where the parameter C_B^s has different time-dependent values at the different locations:

$$C_B^s = \begin{cases} \frac{\Delta_t}{\epsilon_0\epsilon_\infty + Y g_b \Delta_t} & \partial\mho \\ \frac{\Delta_t(1-Y)}{\epsilon_0\epsilon_\infty + \sigma^n_{fe^-} + \epsilon_0\chi_0^{(3)}\alpha|\vec{E}^{s-\frac{1}{2}}|^2 + |\vec{Q}_a^{s-\frac{1}{2}}|} & D - \partial\mho \end{cases}$$

where $\partial\mho$ implicates any region in metals or CPMLs, meanwhile $D - \partial\mho$ represents any computational region filled by a dielectric medium. Finally we have two parameters, g_c and $Y(\vec{r})$, that are defined as follows:

$$g_c = Y\frac{1+g_a}{2}$$

$$Y(\vec{r}) = \begin{cases} 1 & \vec{r} \in \text{Metal} \\ 0 & \vec{r} \notin \text{Metal} \end{cases}$$

Y permits to switch the algorithm from dielectric to metal and vice versa. g_c is a coefficient gathering parameters for convenience in the metal current. Table 1 summarizes the algorithms presented in this section, and Figure 2 sketches the general flow chart. In the next section, we discuss the stability conditions for these algorithms.

Table 1. The list of the new explicit algorithms added to the classical FDTD scheme in order to deal with the nonlinear effects. The current list of algorithms is depicted in the form of a flow chart and it is complemented by the global integration view shown in Figure 2.

Step 0:	Update 6D-Discrete Incident Electric Field Array
	↓
Step 1:	Total field / Scattering field algorithm on $\vec{E}$
	↓
Step 2:	$\lvert\vec{Q}_a^{s+\frac{1}{2}}\rvert = \lvert\vec{Q}_a^{s-\frac{1}{2}}\rvert + \overline{\overline{g}}\left(s\Delta_t - \frac{(k-k_0)\Delta_z}{H_c c_0}\right)\lvert\vec{E}^{s-\frac{1}{2}}\rvert^2$
	↓
Step 3:	$\vec{E}^{s+\frac{1}{2}} = \vec{E}^{s-\frac{1}{2}} + C_B^s\left(\nabla \wedge \vec{H}^s - g_c\vec{J}_{Metal}^{s-\frac{1}{2}}\right)$
	↓
Step 4:	$\vec{J}_{Metal}^{s+\frac{1}{2}} = g_a\vec{J}_{Metal}^{s-\frac{1}{2}} + g_b\left(\vec{E}^{s+\frac{1}{2}} - \vec{E}^{s-\frac{1}{2}}\right)$
	↓
Step 5:	$\lvert\vec{A}^{s+\frac{1}{2}}\rvert = \lvert\vec{A}^{s+\frac{1}{2}}\rvert + \lvert\vec{E}^{s+\frac{1}{2}}\rvert\Delta_t$
	↓
Step 6:	$T_{ci}^{\Sigma_{Time}^{s+\frac{1}{2}}} = T_{ci}^{\Sigma_{Time}^{s+\frac{1}{2}}} + \frac{-\Delta_t\sqrt{E_g}\lvert\vec{E}^{s+\frac{1}{2}}\rvert e^{j\Delta_t\left(s+\frac{1}{2}\right)(E_g+\xi_{ci})}}{(E_g+\xi_{ci})\sqrt{2m_{vc}^*}\left(1-\frac{1}{\Omega}\lvert\vec{A}^{s+\frac{1}{2}}\rvert\right)}\ \forall\xi_{ci} = 0,\ldots,W_f$
	↓
Step 7:	$n_{ph-e^-}^{s+\frac{1}{2}} = N_0\sum_{\xi_{ci}=0}^{W_f}\left\lvert T_{ci}^{\Sigma_{Time}^{s+\frac{1}{2}}}\right\rvert^2$
	↓
Step 8:	$n_{fe^-}^{s+1} = \frac{2-\Delta_t\left(f_r^{s+\frac{1}{2}}-f_a^{s+\frac{1}{2}}-f_\eta^{s+\frac{1}{2}}\right)}{2+\Delta_t\left(f_r^{s+\frac{1}{2}}-f_a^{s+\frac{1}{2}}-f_\eta^{s+\frac{1}{2}}\right)}n_{fe^-}^s + \frac{2f_d^{s+\frac{1}{2}}\Delta_t}{2+\Delta_t\left(f_r^{s+\frac{1}{2}}-f_a^{s+\frac{1}{2}}-f_\eta^{s+\frac{1}{2}}\right)}n_{ph-e^-}^{s+\frac{1}{2}}$
	↓
Step 9:	CPML on Magnetic Field
	↓
Step 10:	Update 6D-Discrete Incident Magnetic Field Array
	↓
Step 11:	Total field / Scattering field algorithm on $\vec{H}$
	↓
Step 12:	$\vec{H}^{s+1} = \vec{H}^n - \frac{\Delta_t}{\mu}\nabla \wedge \vec{E}^{s+\frac{1}{2}}$
	↓
Step 13:	CPML on Magnetic Field
	↓
Step 14:	$\vec{E}^{s-\frac{1}{2}} = \vec{E}^{s+\frac{1}{2}}; \vec{J}_{Metal}^{s-\frac{1}{2}} = \vec{J}_{Metal}^{s+\frac{1}{2}};$ $n_{fe^-}^s = n_{fe^-}^{s+1}; \vec{H}^s = \vec{H}^{s+1}; \lvert\vec{Q}_a^{s-\frac{1}{2}}\rvert = \lvert\vec{Q}_a^{s+\frac{1}{2}}\rvert;$

In Appendix A, we determine the stability conditions for the presented algorithms. Table 2 summarizes the stability conditions of all these effects together or separately.

Table 2. List of stability conditions. Note that $Min[]$ selects the minimum value in a list.

Stability Conditions		
All cases	$\Delta_t \leq \frac{n_i}{c_0}\sqrt{\frac{1}{\Delta_x^2}+\frac{1}{\Delta_y^2}+\frac{1}{\Delta_z^2}}$	
Metal	$\mathcal{R}[a] \geq 0$	$\mathcal{R}[b] \geq 0$
Kerr Raman	$\lvert\widetilde{\vec{E}}_0\rvert < \sqrt{\frac{\epsilon_\infty}{D_f \chi_0^{(3)}}}Min\left[\frac{1}{\sqrt{\alpha}}, \frac{1}{\sqrt{1-\alpha}}\right]$	
Photoionization	$\sigma_{fe^-} \geq 0$	

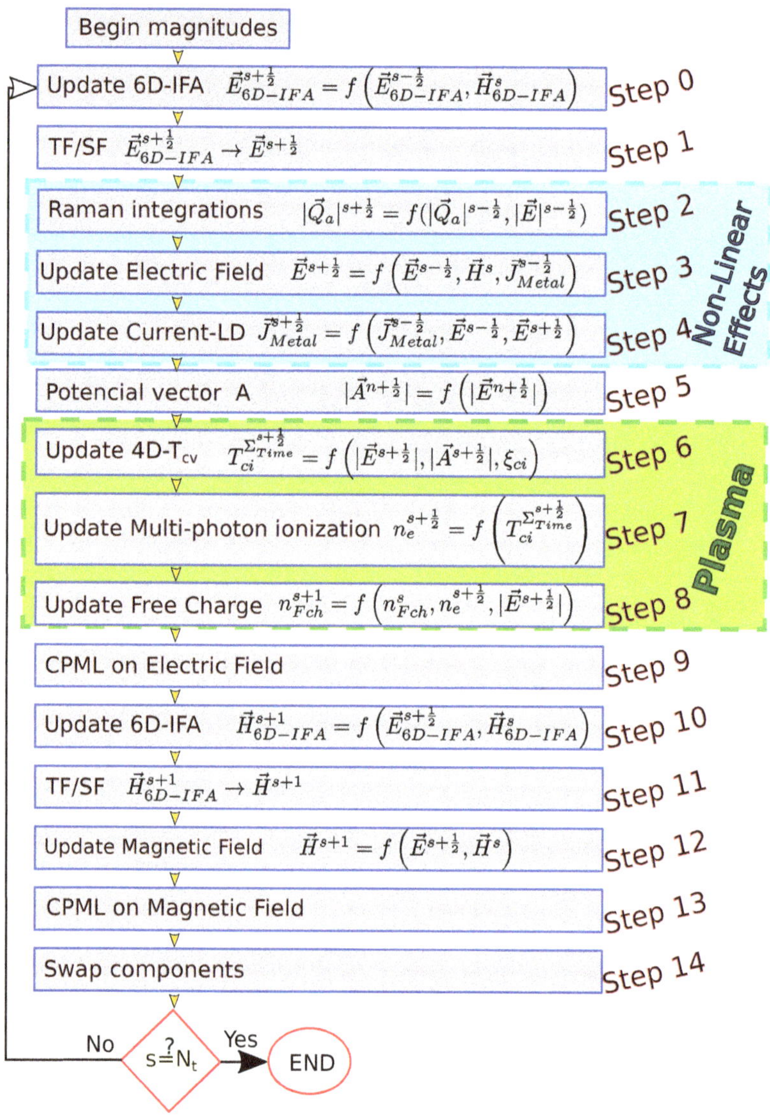

Figure 2. Global flow chart of the proposed GVADE-FDTD algorithm.

4. Performance and Framework

Some approximations made here are not always applicable and, in this section, we will establish the validity framework of the proposed algorithms. In particular, we should consider the two approximations $\int_0^{(s+\frac{1}{2})\Delta_t} \delta(t-\tau)|\vec{E}(\vec{r},\tau)|d\tau \simeq \int_0^{(s-\frac{1}{2})\Delta_t} \delta(t-\tau)|\vec{E}(\vec{r},\tau)|d\tau$ and $|\vec{Q}_a^{s+\frac{1}{2}}| = |\vec{Q}_a^{s-\frac{1}{2}}| + \overline{\overline{g}}((s-nt_0(k))\Delta_t)|\vec{E}^{s-\frac{1}{2}}|^2$. Both expression are derived in [9] under the Born–Oppenheimer approximation [4] where the Kerr non-resonant virtual electronic transitions are in the order of 1fs, so it can be considered an instantaneous event, which is modeled by the Dirac delta function. However, while the time stepping keeps beneath 1 fs, we could model the medium reaction by using a time narrow Gaussian function with a standard deviation of 1 fs. In our simulations, $\Delta_t \sim 0.1$ fs, so we are adding in a fraction of $\frac{1}{10}$ the standard deviation. In this particular situation, the approximation is assumed to be valid. We employ the full explicit form $|\vec{Q}_a^{s+\frac{1}{2}}| = |\vec{Q}_a^{s-\frac{1}{2}}| + \overline{\overline{g}}((s-nt(k))\Delta_t)|\vec{E}^{s-\frac{1}{2}}|^2$ in the place of right implicit form $|\vec{Q}_a^{s+\frac{1}{2}}| = |\vec{Q}_a^{s-\frac{1}{2}}| + \overline{\overline{g}}((s-nt(k))\Delta_t)\left|\frac{\vec{E}^{s+\frac{1}{2}}+\vec{E}^{s-\frac{1}{2}}}{2}\right|^2$ based on the same idea than before, $\Delta_t \ll \tau_1$ and $\Delta_t \ll \tau_2$ so $e^{\frac{-(t+\Delta_t)}{\tau_1}\sin\left(\frac{t+\Delta_t}{\tau_2}\right)} \backsimeq e^{\frac{-(t)}{\tau_1}\sin\left(\frac{t}{\tau_2}\right)}$. This expression is valid for all instant $\forall t$. Hence, we fix the limit for this full explicit approach in the relation $\Delta_t < 1$ fs. The advantage of a fully explicit approach concerns the simulation time. Due to the updating connection in an implicit scheme, most of the algorithms should be re-updating to advance a time step. In classical FDTD, it is usually fixing the limits of the index in all nested loops. In a significant computational domain, for instance, 1.2 G cells, the updating demands 30 s. If the algorithm has to be repeated at least three times, the simulation should need around three times the the implicit scheme. Figure 3 shows the number of simulation status updates on the abscissa axis and on the ordinate axis the time needed in order to update these simulation states. Each step or update corresponds to executing the 14 algorithms shown by Figure 2. In addition to these algorithms that are necessary to perform a system status update, two numerical processes are employed. One exports simulation data or results (every 40 updates) and the other exports the complete simulation status (every 500 updates). The latter is done for two reasons. The first is to recover a state of a simulation in case of computer collapse (for example a failure in the computer electric power source). The second reason concerns the alteration of the computational domain (electromagnetic properties or/and size of the computational domain) without the need to restart the simulation from the first time stepping. In Appendix B the Table A1 lists the computational resources employed in the simulations. Under these conditions we reach a performance which can be summarized by the computational burdens and some performance parameters. Those relies on a computational domain size of 121.5 GBs in RAM, a maximum number of Yee's cell charged and computed of 1.2 Giga cells, a minimum number of Yee's cell computed of 13 Mega cells, a speed at minimum and maximum load of 6.5 Mcells/s and 38 Mcells/s respectability, a probe exportation time of 1.47 s and a computational domain backup time of 8.76 s. Therefore, Figure 3 depicts the time performance for a simulation carried out with dynamics nested loops. In our scheme, the size of the nested loops is provided by the information created. We know that our implementation violates the second rule of NASA's Ten-Rules for Developing Safety Critical Code [19]. In general, it is a good practice to arrange the loop indexes in the simulation beginning and keep these ranges. By doing this, we fix, from the start, the nested loop size. Nevertheless, the employment of a dynamic nested loop saves simulation time. Figure 3 depicts that we save more than forty hours. In a simulation with the same characteristics and static nested loop, we should have a constant duration per time step updating of 32 seconds which is equal to the asymptotic value in Figure 3. The total simulation duration is the accumulated time or area under the plot schematized in Figure 3. Hence, the triangle in gray color is the time saved by means of dynamic nested loops.

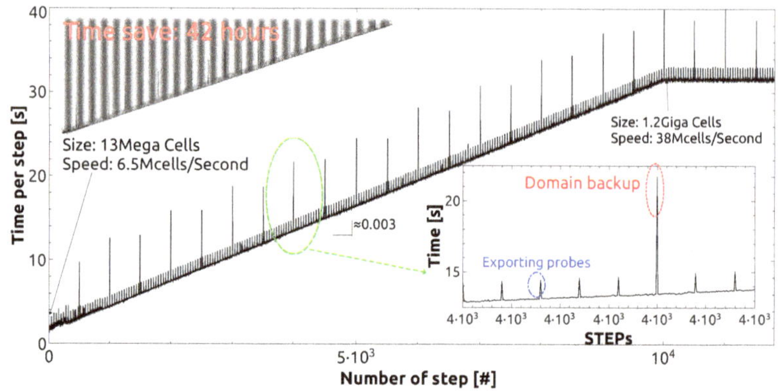

Figure 3. The figure shows the time per step in the simulation updating in a computational domain with dynamics limits (index k is a dynamical magnitude). Furthermore, the number of threads in the palatalization process is dynamics.

Finally, we close this section with some consideration on Equation (3). This equation is a derivation from the charge/electrons continuity equation [20], which has the general form $\frac{\partial\rho(\vec{r},t)}{\partial t} = q(G(\vec{r},t) - R(\vec{r},t)) - \nabla \cdot \vec{J}(\vec{r},t)$, where we neglect the term $\nabla \cdot \vec{J}(\vec{r},t) \ll qn_{fe^-}(\vec{r},t)$. We know from theory that $\rho(\vec{r},t) = qn_{fe^-}(\vec{r},t)$ and $\vec{J}(\vec{r},t) = q\mu_{fe^-}n_{fe^-}(\vec{r},t)\vec{E}(\vec{r},t)$. The key is the extremely low mobility $\mu_{fe^-} \sim 10^{-22}$, which is deduced from [21]. We can consider a highly intense laser beam with $|\vec{E}(\vec{r},t)|^2 \sim 10^{13}$, which induces a free electron concentration in fused silica of $n_{fe^-}(\vec{r},t) \sim 10^{26}$ and compare with $\frac{\nabla\cdot\vec{J}(\vec{r},t)}{q} \sim \frac{\frac{10^{-22}\times q 10^{26}\times 10^{13/2}}{10^{-7}}}{q} \sim 10^{18}$. Therefore, we can conclude that $n_{fe^-}(\vec{r},t) \sim 10^{8}\frac{\nabla\cdot\vec{J}(\vec{r},t)}{q}$ and Equation (3) is a valid approximation in the present context.

5. Results

An efficient technique to compute scattered fields in the context of FDTD modeling is the total-field/scattered-field (TF/SF) incident wave source, which is employed by most all current commercial FDTD solvers [2]. Fundamentally, the TF/SF technique is an application of the well-known electromagnetic field equivalence principle [22]. By this principle, the original incident wave of infinite extent and arbitrary propagation direction, polarization, and time-waveform is replaced by electric and magnetic current sources appropriately defined on a finite closed surface, called the Huygens surface [23], containing the object of interest. The reformulated problem confines the incident illumination to a compact total-field region and provides a finite scattered-field region external to the total-field region that is terminated by an absorbing boundary condition (ABC) to simulate the FDTD grid extending to infinity. In particular, we use convolutional perfectly matched layers, a technique already used in other works [24]. In this work, we have used the finite-difference time-domain discrete plane wave technique (FDTD-DPW), which is a numerical approach based on TF/SF technique that allows one to propagate plane waves quasi-perfectly isolated [25]. This means a propagation in the total field domain without reflections to the scattered field domain on the order of machine precision (∼300 dB) [26]. This technique is valid for any angle of propagation, for any grid cell aspect ratio and even for nonuniform grids [27]. Figure 4 illustrates the total field region where there are two important elements, an axicon lens that generates the Bessel beam from the initial plane waves and a block of fused silica. In this domain, the incident plane wave is generated by the FDTD-DPW scheme.

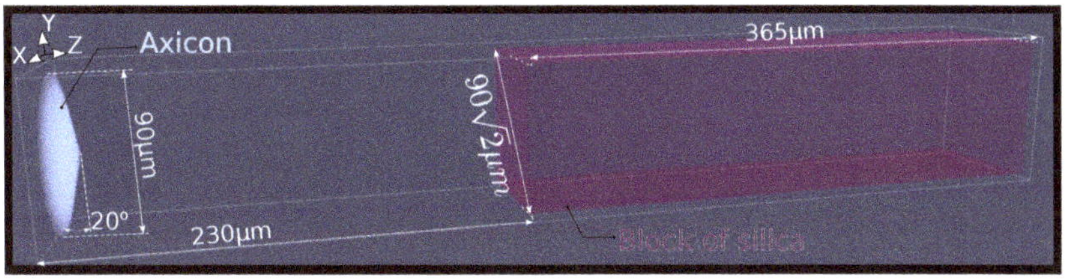

Figure 4. Computational domain.

The calculation domain contains a block of fused silica glass where the nonlinear effects apply. The interaction between the incident plane wave and the axicon lens results in a Bessel beam, which impinges the SiO_2 block. We assume a pulsed laser source of wavelength (800 ± 11) nm and a time full width at half maximum of $\sigma_t = 40$ fs. The pulsed incident time-waveform is modeled by $h(t) = A(t)\text{sech}^2\left(\frac{t-t_0}{\sigma_t}\right)\sin(t\omega)$, where $A(t) = \left(\frac{\tanh(t\omega)+1}{2}\right)^{\varsigma}$ with $\varsigma \geq 1$, is an appropriate way to avoid numerical instabilities in the simulation beginning. The incident field six-dimensional array (IFA) that feeds the FDTD-DPW is quite simple $\vec{E}(\vec{r}_{IFA,t}) = E_0 h(t)\hat{x}$. In addition, there are other parameters that characterized the media properties and are summarize in Table 3.

Table 3. Media main parameters.

Parameters				
Name	**Symbol**	**Value**	**Unit**	**Ref.**
Gap energy	E_g	9	eV	[28]
Third-order elec-tric susceptibility	$\chi_0^{(3)}$	10^{-22}	$\frac{m^2}{V^2}$	[28]
Raman sinusoidal time	τ_1	12.2	fs	[29]
Raman response fraction	$1-\alpha$	0.18	-	[29]
Electron recombi-nation time	τ_r	150	fs	[10]
Mobility	μ_{fe^-}	1.13×10^{-22}	$\frac{m^2}{Vs}$	[21]
Density of valence electrons	N_0	10^{22}	cm^{-3}	[10]
Electron mass reduced particle	m^*_{vc}	0.5	m_0	[10]
Work function	W_f	5	eV	[28]
Avalanche ioniza-tion coefficient	Θ	10^{-3}	$\frac{m^2}{Ws}$	[28]
Raman decay time	τ_2	32	fs	[29]
Kerr response fraction	α	0.82	-	[29]
Laser characteristic time for electrons production	τ_d	3	ps	[10]
Impact ionization efficiency	η	0.01	-	[28]
Fitting parameter for the spatial expansion of the valence wave function	Ω	0.8	-	[10]
Metal parameters	a b	$19.2 + j2.7$ $1.3 - j3.9$	PHz PHz	[15]
Axicon refractive index	n_a	1.45	-	[30]

By using the developed irradiation strategy, the computational domain elements and their properties as well as the algorithm we carried out, the obtained results can be divided into three groups:

- Low power regime: $E_0 = 10^4 \frac{\mathrm{V}}{\mathrm{m}}$ and $I = 2.46 \times 10^{14} \frac{\mathrm{W}}{\mathrm{m}^2}$.
- Critical power regime: $E_0 = 3.5 \times 10^5 \frac{\mathrm{V}}{\mathrm{m}}$ and $I = 8.61 \times 10^{16} \frac{\mathrm{W}}{\mathrm{m}^2}$.
- High power regime: $E_0 = 10^6 \frac{\mathrm{V}}{\mathrm{m}}$ and $I = 2.45 \times 10^{18} \frac{\mathrm{W}}{\mathrm{m}^2}$.

The power intensity $I = Max(\widetilde{I}(\vec{r}, t))$ is calculated as the maximum power that flows through a section of σ_w^2, being σ_w the pulsed laser transverse waist, σ_ℓ the pulsed length in the propagation direction and z_p the central pulse location at the calculation time. Hence, $\widetilde{I}(\vec{r}, t) = \iint_{-\sigma_w}^{\sigma_w} dxdy \int_{z_p-\sigma_\ell}^{z_p+\sigma_\ell} dz\hat{z} \cdot \frac{\vec{S}(\vec{r},t)}{\sigma_w^2}$.

To validate the algorithm, which includes nonlinear effects during light propagation, we have investigated the features of the propagation regime depending on the laser intensity. Figure 5a,b redraw the distribution evolution of the pulsed square electric field module for simulation performed for different laser intensity sources. Figure 5a plots the iso-slices that depicts $|\vec{E}|^2$ inside the fused silica bulk for high laser power in the system. From top to bottom, the time evolution sequence reveals the defocusing process of the Bessel beam, which ends by forming some filamentation structures in the light distribution. At the time $t = 5720, \Delta_t = 1.52$ ps, the pulsed square electric field module begins to penetrate the interface air-fused silica. All successive time plots correspond to a propagation inside the fused silica block. Figure 5b illustrates, for the same time moments, but a higher initial power source, the square electric field module during the propagation inside SiO_2 material.

We complement the picture with a simulation carried out in the medium power regime. From the instant of time, $t_1 = 1.67$ ps to the instant $t_{18} = 2.56$ ps, Figure 5c illustrates the antagonistic trade-off between the focusing and the defocusing process experienced by the Bessel beam along with its travel through the bulk fused silica target. Figure 6b shows the history of the power density at the critical threshold separating medium and high power regimes, along with a few steps of time. In this figure we can identify the interface air-fused silica due to the rings which record the history of a stationary wave formed after the interaction. We can observe a strong focusing process in the entrance of the fused silica block followed by an alternation of refocusing and the defocusing events induced and governed by the photoionization and Kerr effect competition, and we know that these are not diffraction artifacts because in the air, at the same laser intensity, we do not observe them. At this stage, we have to give a clear definition of the pulsed power density and the power history. The magnitude of pulsed power density is defined as $\nabla \cdot \vec{S}(\vec{r}, t)$, where $\vec{S}(\vec{r}, t) = \vec{E}(\vec{r}, t) \wedge \vec{H}(\vec{r}, t)$ is the Poynting vector [31]. Figure 6a sketches the pulsed power density. From this magnitude $\nabla \cdot \vec{S}(\vec{r}, t)$, we can calculate the delivered pulsed electromagnetic density of energy $\int_0^{t'} \nabla \cdot \vec{S}(\vec{r}, t)dt$. Besides this, it is interesting to consider the range of values in Figure 6a. We can identify a negative power density that requires clarification. As the Poynting vector represents the flux of power through a given surface, the divergence of this magnitude represents the power density that comes into/out a particular point in a given region. In particular, in FDTD calculations, we are describing the power flux through the computational nodes that discretized the computational domain. In the density power accounting, the reference system is placed inside the cell that encloses the nodes. In this way, a negative power density, in a particular instance of time, in a local region, means a local sink of power $\nabla \cdot \vec{S}(\vec{r}, t) < 0$ in that region, and vice versa, we have a local source of power as $\nabla \cdot \vec{S}(\vec{r}, t) > 0$. These oscillations in the density power magnitude indicate a redistribution that is able to support the pulsed density power and the pulsed density energy traveling.

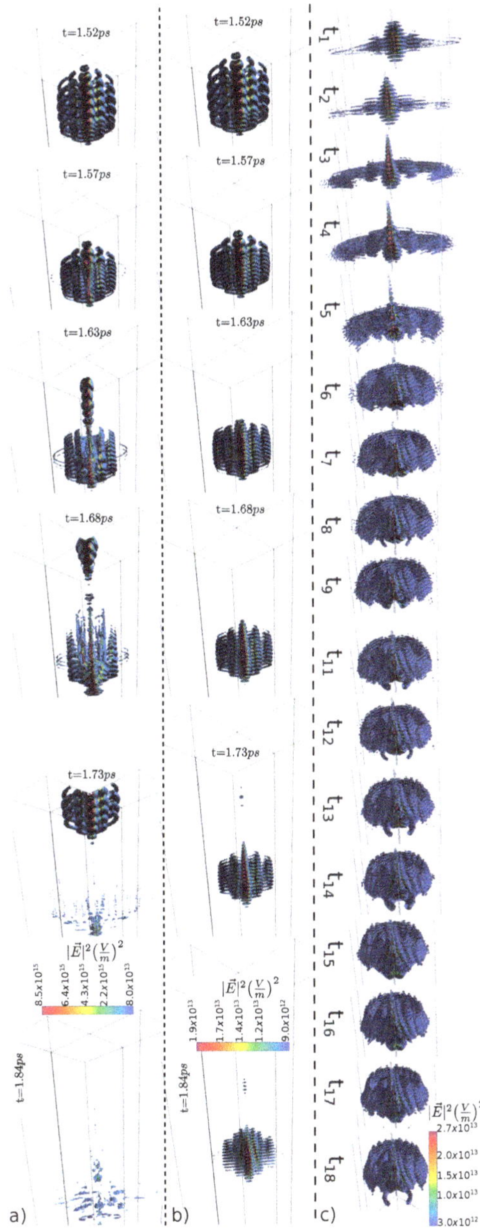

Figure 5. Propagation of the pulsed beam given by a time sequence of instants which has been plotted before reaching the theoretical critical power, with an energy per pulse of 2.8 μJ allowing to reach an electron density of 10^{27} m^{-3}, the beam splits, enclosing the charge and defocusing the beam. (**a**) Several instants of the pulsed square electromagnetic field when the power is beyond the critical power (High power regime). (**b**) Pulse propagation sequences during the travel inside the fused silica bulk for low power regime. (**c**) Sequence of frames that sketches the pulse along during its travel inside the fused silica bulk from the instance $t_1 = 6290\Delta_t$ to $t_{18} = 9620\Delta_t$ (where $\Delta_t = 2.65765 \times 10^{-16}$ s and Courant number 0.69). The difference between two time-steps is $200\Delta_t$ in the sequence. The power of this sequence is in the threshold between low and high power regimes (critical power).

Turning to the matter at hand, we define the history of the power density by applying the absolute value to the divergence of the Poynting vector $|\nabla \cdot \vec{S}(\vec{r},t)|$.

The bi-dimensional longitudinal cut of the history of the power density depicted in Figure 6b permits one to identify the air–fused silica interface. There, in that interface, we can appreciate the reflected beam and the strong focusing in the entrance due to the Kerr effect. We assume the same refractive index for both the axicon lens and the fused silica block in which the wave Bessel beam impinges.

Although we assume nonlinear media, in the simulation results presented up to this point in the manuscript, we have considered that the fused silica is isotropic and homogeneous. However, in order to study the effect of micro- and nanopores in the fused silica, 650 regions or voids have been introduced that range in size from 200 to 650 nanometers in radius. These spherical embedded cavities, that are plotted in Figure 7a, emulate a nanoporous fused silica bulk. The traditional method of linear combined congruence random number generators has been used to assign the size and location of these pores of air/void particles. The linear congruential generator yields a sequence of pseudo-randomized numbers calculated with a discontinuous piecewise linear equation. The most important thing about the method is that, under the same seeds, the same random number distributions will be obtained, which allows us to reproduce both size distributions and particle locations. This will be useful to evaluate the effects resulting from another kind of media, such as metal particles, which will allow us to make a point-to-point comparison and infer conclusions from the results. In this setup, Figure 7b shows the history of the power in the region where the particles are located. The simulation is stopped when the nonlinear Bessel beam exits the porous fused silica region. Looking at Figure 7c, it appears that a correlation can be made between the distribution of photoionized electrons and the laser power density, in full consistency with our model. Figure 7c shows a history of the density of the ionized electrons as a consequence of the Bessel beam, which travels through the fused silica block. This result leads us to conjecture that both quantities look similar in terms of spatial structure.

It is relevant to study the distribution or history of power in a non-homogeneous dielectric, as is the present case. If we observe Figure 7b in detail, we see that the history of power reveals the history of field intensity as it passes through the nanoporous medium of SiO_2. Due to a simple matter of impedance, the wavefront of the pulse is redistributed, avoiding entering inside the voids. As we can see in the longitudinal section shown by Figure 7b, the power vanishes inside the voids and the maximum power is not found at the interface fused silica-void. The maximum is reached in the glass medium between the pores. There is a visual explanation for this behavior. In some lenses, an anti-coating multi-layer that matches the impedance of the lens with the surrounding media (in general, glasses could be done for water or air), is coupled in order to remove the reflections. This gradient in the refractive index is necessary to facilitate the flowing energy from the lens to the environment. In the porous fused silica, there is an abrupt change in the refractive index that explains our result.

There is a Supplementary Material consisting in a video animation (see the link https://youtu.be/nvGdtxQ9E8o) that shows dynamically the ultrafast pulsed Bessel beam, depicted by employing the pulsed electric field module, that crosses the interface air-fused silica and travels through a fused silica block with air nanovoids.

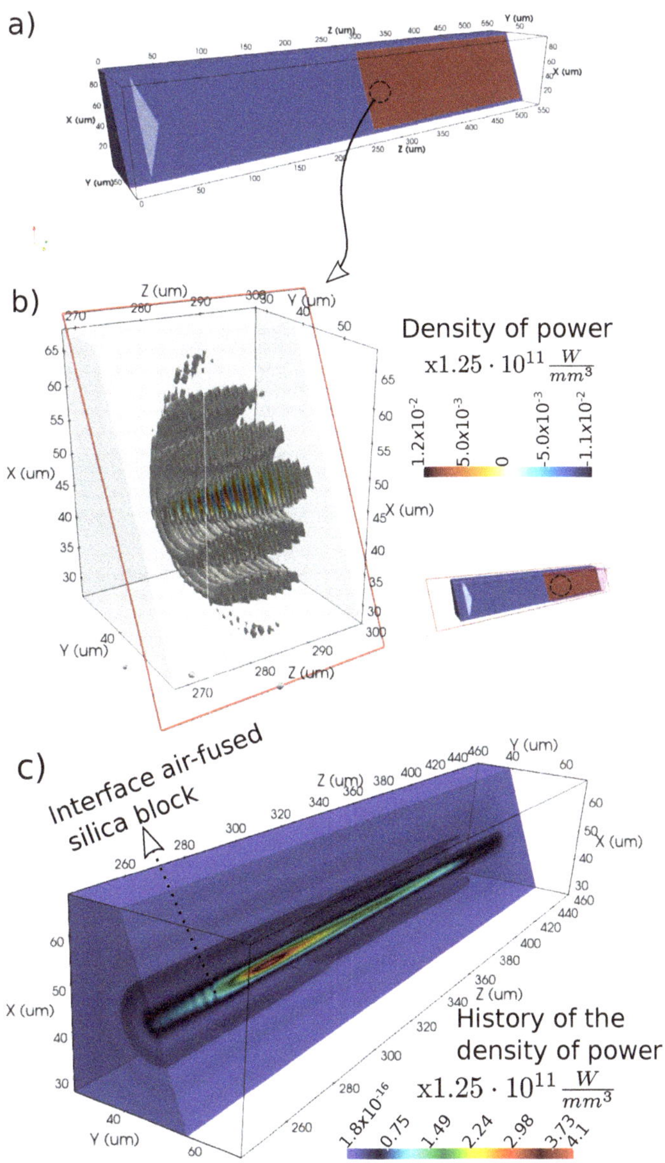

Figure 6. (**a**) Subsample taken from the computational domain in which we plot the pulsed density of power and the history of the power density. (**b**) Power density power plotted by means of iso-slices for a pulsed Bessel beam in power regime close to the critical value. the selected time corresponds to the Bessel beam entrance inside the fused silica block. We consider all nonlinear effects in this simulation. (**c**) History of the power density for a pulsed Bessel beam. At the interface between the fused silica bulk and the air a modulation is observed in the history of power density.

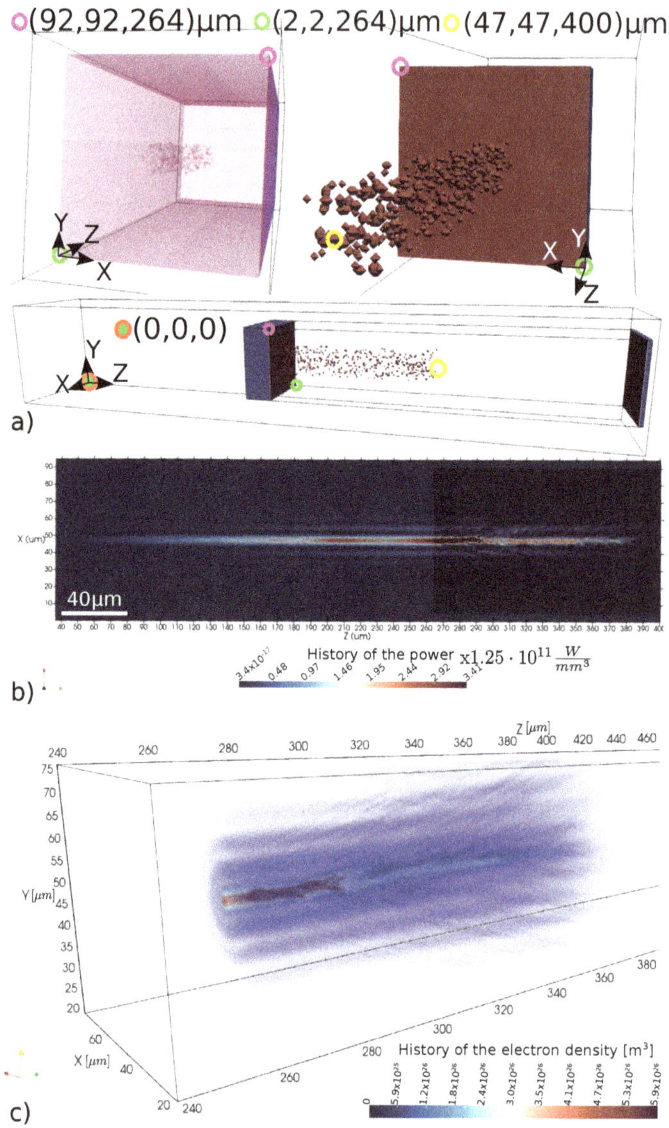

Figure 7. (**a**) Computational domain including clusters of spherical nanovoids, randomly located inside the fused silica bulk. (**b**) Power history of the pulsed Bessel beam propagating inside the nanoporous block. (**c**) History of the photo-ionized electrons density in the complete domain.

As stated above, we have followed identical reasoning and repeat the simulation, this time replacing nanovoids with metal particles. The results are shown in Figure 8. Figure 8a illustrates the power history under the same intensity of light. Along the photoexcitation path, the maximum power occurs around the metal– fused silica interface, enhancing and localizing the light coupling to the nanocomposite material. Even if the non-diffractive behavior is still present, it is less observable, as the beam is strongly attenuated by metal absorption. Moreover, the arrangement of the higher region of light absorption follows the randomly distributed composition of the metallic sphere. This indicates that contrary to dielectric–void interfaces, the dielectric-metal region fosters local effects, preventing any collective response of the medium. This statement may differ if the particles are judiciously defined in terms of concentration and size in order to enhance resonance effects as plasmonic modes. In Figure 9, we observe a pattern in the electron photoionized history. In particular, Figure 9a depicts a weak structure of parallel planes, which is more intense in some regions. Figure 9b shows a stronger pattern. The present results, sketched by Figure 9a, reveal that ultrashort laser pulse interacting with distributed performed scattering nanovoids induces local field enhancement that results in enhanced nonlinear absorption and localized optical breakdown. At the void-SiO_2 interface, the multiphotonic ionization process transforms the material into a thin layer of absorbing plasma with metallic-like properties. The light coupling initiates the growth of nanoplates of enhanced energy deposition, perpendicular to laser propagation direction, but parallel to the polarization. This kind of arranged structure exhibiting a periodicity approximately the wavelength of light in fused silica was commonly experimentally reported in conditions of tightly focused and high repetition rate [32,33]. They directly emerge from the spatial coherence of the waves scattered from the inhomogeneous material in the plane perpendicular to the beam propagation direction. Note the relatively used large nanovoids associated to the grid resolution do not allow to discriminate nanogratings of higher frequency perpendicular to the electric field. For propagation inside laser-induced nanoporous media, the fused silica glass remains underdense, below the critical plasma concentration, and the standing wave effect remains weak. However, the other simulation conditions, portrayed in Figure 9b), clearly indicate that for metallic nanoparticles, local field enhancement occurs near the metal-dielectric interface [34]. Upon electron–matrix energy coupling and void deformation, the light pattern may result in elongated nanostructures [35]. This shows that nanoplasmonic behavior dominates and opens the route for advanced applications as embedded micro-reflectors for a stronger near-field interaction regime.

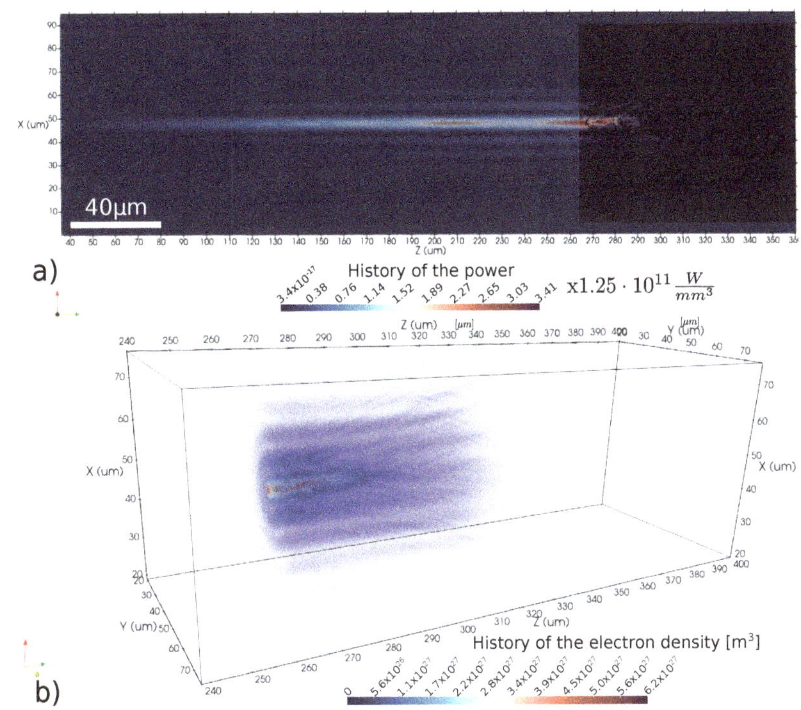

Figure 8. (**a**) Power history of the pulsed Bessel beam propagating inside the fused silica block in which we place a cluster of spherical metal particles. (**b**) History of the photo-ionized electrons density in the complete domain.

Propagation and scattering through sphere inclusions with different optical properties demonstrate that the current algorithms are stable and can model efficiently nonlinear media with sharp dielectric interfaces among them. Moreover, nonlinear metal–dielectric interface investigations are far reaching for technology-based photonics. For instance, in a wider frequency spectrum, resonators, metasurfaces with designed cavities, metal–dielectric waveguides or even coating optical fibers are composed of metal–dielectric interfaces that may involve simultaneously nonlinear mechanisms of different nature. If we imagine for a sake of simplicity an harmonic electromagnetic pulse, which is guided, for instance, inside of a optical fiber, we should solve two coupled inhomogeneous Helmholtz's equations in the frequency domain $\nabla^2\vec{\tilde{E}}(\vec{r}) + \omega^2\mu\epsilon(\vec{r})\vec{\tilde{E}}(\vec{r}) = -\nabla\left(\vec{\tilde{E}}(\vec{r})\cdot\nabla(\ln(\epsilon(\vec{r})))\right)$ and $\nabla^2\vec{\tilde{H}}(\vec{r}) + \omega^2\mu\epsilon(\vec{r})\vec{\tilde{H}}(\vec{r})(\vec{r}) = (\nabla\wedge\vec{\tilde{H}}(\vec{r}))\wedge\nabla\ln(\epsilon(\vec{r}))$. In these equations, the refractive index $n(\vec{r}) = \sqrt{\epsilon_r(\vec{r})\mu_r}$, which is a dynamic magnitude along the pulsed beam propagation or simply a nonlinear magnitude, has to be taken into consideration. In this way, the algorithms present here can be used to deal with this kind of problems.

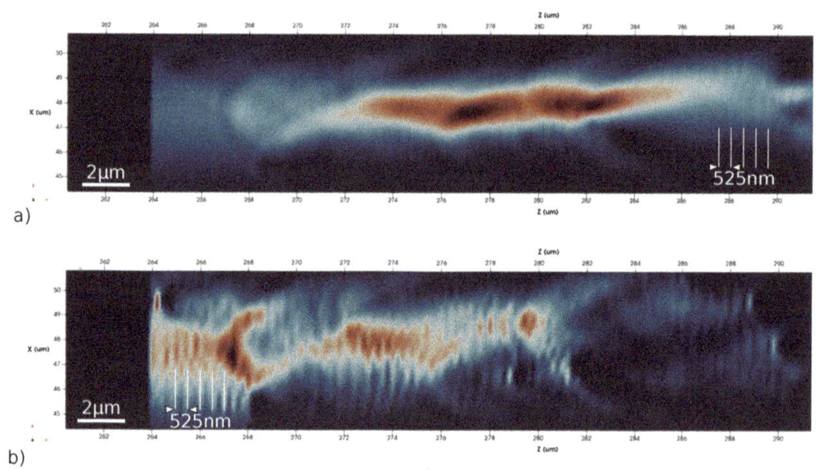

Figure 9. Pattern observed in the history of the photo-ionized electron history for (**a**) voids and (**b**) metal particles.

6. Conclusions

In this work, we have proposed a numerical approach for nonlinear laser beam propagation based on the time-domain discretization of Maxwell equations in a fully explicit scheme, including four nonlinear optical effects. The developed algorithms have been benchmarked in three different ways. First, we have studied the algorithm's stability conditions by combining the von Neumann method with the Routh–Hurwitz criterion. Second, we have established the physical framework for the algorithms, with special attention given to the approximation carried out in the algorithm derivation. Third, we have employed the algorithms in order to determine the transient optical properties of a photoexcited media when a laser beam is focused inside a fused silica bulk. In particular, we have considered a Bessel beam as a free-diffraction beam due to its stable and rebuildable-like energy distribution along with electromagnetic pulse travel path. This non-diffractive behavior has been observed numerically by tracking the power/energy history. This has been determined by evaluating the absolute value of the power $|\nabla\cdot\vec{S}(\vec{r},t)|$ and the density energy history as $\int_0^{t'}|\nabla\cdot\vec{S}(\vec{r},t)|dt$. The main advantage of the present method consists of its remarkable numerical performance (see Figure 3), that together with the tolerant stability conditions found out and the accuracy results, under an also permissive physical framework, made it valid for a wide range of applications (such as ultrafast laser processing which has been extensively applied to fabricate 3D photonic devices based on embedded optical waveguides inscribed in glass materials by inducing permanent refractive index changes in the focal volumes [36], optical couplers and splitters [37], volume Bragg gratings [38], or even diffractive lenses [39] which have been efficiently inscribed by refractive-index changes during light propagation) in which optical/electromagnetic nonlinear effects have to be taken into consideration. Finally, our approach is generalizable to any nonlinear electromagnetic problem with dielectric–metal interfaces. The proposed work elucidates the simultaneous treatment of nonlinear effects and provides new routes toward the simulation of light–matter interaction with sub-wavelength features for the optimization of guided structures or loss-free metasurfaces.

Supplementary Materials: The following are available at https://youtu.be/nvGdtxQ9E8o.

Author Contributions: Conceptualization, E.M. and J.-P.C.; methodology, E.M.; programming, E.M.; validation, E.M.; formal analysis, E.M. and J.-P.C.; writing—original draft preparation, E.M. and J.-P.C.; writing—review and editing, all authors; visualization, E.M.; supervision, J.-P.C. All authors have read and agreed to the published version of the manuscript.

Funding: This work was supported by the FUI project 3D HYBRIDE financed by the French Single Interministerial Fund and by the LABEX MANUTECH-SISE (ANR-10-LABX-0075) of the Universite de Lyon, within the program "Investissements d'Avenir" (ANR-11-IDEX-0007) and the INTRALAS project (ANR-19-CE30-0036) operated by the French National Research Agency (ANR).

Institutional Review Board Statement: Not applicable.

Informed Consent Statement: Not applicable.

Data Availability Statement: All data is provided in the manuscript.

Acknowledgments: The authors would like to thanks the free use of some tools under General Public License, that have facilitated this work: Gcc, Gfortran, Guile-OpenGL, FreeCAD, Paraview, Inkscape, SciDaVis, and Science Eclipse Parallel Tools Platform.

Conflicts of Interest: The authors declare no conflict of interest.

Appendix A. Stability

In this appendix, we study the stability of the algorithm by means of the combination of the von Neumann method with the Routh–Hurwitz criterion [40]. The von Neumann method employs a Fourier series expansion of the error at the mesh nodes at a given time instant $t = s\Delta_t$. Due to linearity, only a single term of this expansion needs to be considered, i.e.,

$$\vec{E}^s|_{i,j,k} = \tilde{\vec{E}}_0 y^s e^{j\left(i\Delta_x\tilde{k}_x + j\Delta_y\tilde{k}_y + k\Delta_z\tilde{k}_z\right)} \tag{A1}$$

$$\vec{D}^s|_{i,j,k} = \tilde{\vec{D}}_0 y^s e^{j\left(i\Delta_x\tilde{k}_x + j\Delta_y\tilde{k}_y + k\Delta_z\tilde{k}_z\right)} \tag{A2}$$

where $\tilde{\vec{E}}_0$ and $\tilde{\vec{D}}_0$ are complex amplitudes, indexes i, j, k denote the position of the nodes in the mesh, Δ_x, Δ_y, Δ_z are the sizes of the discretization cell, and $\tilde{k}_x$, $\tilde{k}_y$ and $\tilde{k}_z$ are the wavenumbers of the discrete modes. The factor y in (A1) and (A2) is a complex variable, often called the amplification factor, which gives the growth of the error in a time updating. To ensure that a finite-difference algorithm is stable, the error must not grow as the time increases, thus, the condition $|y| \leq 1$ must be satisfied. The stability condition for a particular FDTD scheme leads to a characteristic polynomial in y:

$$\Phi(y) = \sum_{\ell=0}^{N_p} a_\ell y^{N_p-\ell} \tag{A3}$$

Assuming that y_ℓ are the roots of $\Phi(y)$, the condition for stability implicates that the roots y_ℓ must lie inside or on the unit circle in the Y-plane. In this stability purpose, the

Routh–Hurwitz criterion provides a method that ensures the roots y_ℓ on the unit circle in the Y-plane. It establishes that the polynomial $\Phi(r) = \sum_{\ell=0}^{N_p} b_\ell y^{N_p-\ell}$ with constant real coefficients b_ℓ has no roots in the right-half of the r-plane if all the entries of the first column of the Routh table are non-negative quantities. In [40], there is a error that makes it impossible to build the Routh table, so we explain here how to build up the Routh table. The first and second rows of the coefficients correspond to even and odd powers of r, respectively. For simplicity in the notation, and without loss of generality, it is usual assuming N_p to be an even number, therefore $c_{1,\ell} = b_{2\ell}$, with $\ell = 0, 1, 2, \ldots, \frac{N_p}{2}$ and $c_{2,\ell} = b_{2\ell+1}$, with $\ell = 0, 1, 2, \ldots, \frac{N_p}{2} - 1$. The remaining entries of the table are obtained by using the following expression $c_{m,\ell} = \frac{c_{m-2,\ell+1}c_{m-1,0} - c_{m-2,0}c_{m-1,\ell+1}}{c_{m-1,0}}$.

The Routh–Hurwitz criterion can be used to determine if any root of the stability polynomial $\Phi(y)$ lies outside the unit circle in the y-plane. To this end, the bilinear transformation $y = \frac{r+1}{r-1}$ is carried out in $\Phi(y)$. This transformation maps the exterior of the unit circle in the y-plane onto the right-half of the r-plane. To satisfy the von Neumann stability condition for a different scheme, as a function of the parameters of interest, all the entries of the first column of the Routh table are forced to be non-negative quantities. Nevertheless, before beginning the search for stability conditions, some definitions are necessary. We define the following operators applied to Ξ^s: $\bar{\partial}_t[\Xi^s] = \frac{\Xi^{s+\frac{1}{2}} - \Xi^{s-\frac{1}{2}}}{\Delta_t}$, $\bar{\Delta}[\Xi^s] = \frac{\Xi^{s+\frac{1}{2}} + \Xi^{s-\frac{1}{2}}}{2}$, $\bar{\partial}_x[\Xi^s] = 2j\sin\left(\frac{\tilde{k}_x\Delta_x}{2}\right)\Xi^s$, $\bar{\partial}_y[\Xi^s] = 2j\sin\left(\frac{\tilde{k}_y\Delta_y}{2}\right)\Xi^s$, and $\bar{\partial}_z[\Xi^s] = 2j\sin\left(\frac{\tilde{k}_z\Delta_z}{2}\right)\Xi^s$, where Ξ^s could be for instance $\vec{E}^s$ or $\vec{D}^s$.

Appendix A.1. Stability Condition on Metal

For dielectric–metal interface, we deal with the wave equation $\mu\frac{\partial^2\vec{D}(\vec{r},t)}{\partial t^2} - \nabla^2\vec{E}(\vec{r},t) = \vec{0}$ in dielectric and the differential equation for the metal writes $(b + \frac{\partial}{\partial t})\vec{D}(\vec{r},t) = \epsilon_0\left(b\epsilon_\infty + a + \epsilon_\infty\frac{\partial}{\partial t}\right)\vec{E}(\vec{r},t)$. After discretization, the wave equation is rewritten as [41],

$$\mu\frac{\bar{\partial}_t}{\Delta_t^2}\vec{D}^s - \left(\frac{\bar{\partial}_x^2}{\Delta_x^3} + \frac{\bar{\partial}_y^2}{\Delta_y^3} + \frac{\bar{\partial}_z^2}{\Delta_z^3}\right)\vec{E}^s = \vec{0}$$

and the auxiliary differential equation for the metal,

$$(b\bar{\Delta} + \bar{\partial}_t)\vec{D}^s = \epsilon_0\left((b\epsilon_\infty + a)\bar{\Delta} + \epsilon_\infty\bar{\partial}_t\right)\vec{E}^s.$$

Now, we replace the electric field $\vec{E}^s$ and the electric displacement field $\vec{D}^s$ in the last two equations by the Equations (A1) and (A2). Hence, from the wave equation we obtain:

$$(y-1)^2\tilde{\vec{D}}_0 + 4y\epsilon_\infty\epsilon_0\nu^2\tilde{\vec{E}}_0 = 0$$

and from the differential equation for the metal:

$$(\Delta_t b(y+1) + 2(y-1))\tilde{\vec{D}}_0 = \Delta_t\epsilon_0((b\epsilon_\infty + a)(y+1) + 2\epsilon_0\epsilon_\infty(y-1))\tilde{\vec{E}}_0$$

where $\nu^2 = \left(\frac{\Delta_t}{\sqrt{\mu\epsilon_\infty\epsilon_0}}\right)^2 \sum_{\wp}^{x,y,z} \frac{\sin^2\left(\frac{\tilde{k}_\wp\Delta_\wp^2}{2}\right)}{\Delta_\wp^2}$. By combining both polynomials we determine the characteristic polynomial $\Phi(y) = a\Delta_t(y+1)(y-1)^2 + \epsilon_\infty(y(4\nu^2 + y - 2) + 1)(b\Delta_t(y+1) + 2(y-1))$. In order to consider the Routh–Hurwitz criterion we apply the bilinear transformation $y = \frac{r+1}{r-1}$ and determine the characteristic polynomial $\Phi(r) = b\Delta_t\epsilon_\infty\nu^2r^3 + 2\epsilon_\infty\nu^2r^2 + \Delta_t(a + b\epsilon_\infty(1-\nu^2))r + 2\epsilon_\infty(1-\nu^2)$. The first column in the Routh table, which should be positive, are $c_{1,0} = b\Delta_t\epsilon_\infty\nu^2$, $c_{2,0} = 2\epsilon_\infty\nu^2$, $c_{3,0} = a\Delta_t$ and $c_{4,0} = 2\epsilon_\infty(1-\nu^2)$. From $c_{1,0}$ and $c_{3,0}$ we find out that $\mathcal{R}[a] \geq 0$ and $\mathcal{R}[b] \geq 0$. The condition over $c_{4,0} \geq 0$

means that $1 - \nu^2 \geq 0$. This is the classic stability condition based on single-cubic FDTD [42] where $\Delta_t \leq \frac{n_i}{c_0}\sqrt{\frac{1}{\Delta_x^2} + \frac{1}{\Delta_y^2} + \frac{1}{\Delta_z^2}}$, where c_0 is the speed of light in the vacuum and n_i the refractive index of the media.

Appendix A.2. Stability Condition on Kerr Effect

In addition to the wave equation, we have to deal with the differential equation associated to the Kerr effect $\frac{\partial}{\partial t}\vec{D}(\vec{r},t) = \epsilon_0\epsilon_\infty\frac{\partial}{\partial t}\vec{E}(\vec{r},t) + \frac{\partial}{\partial t}\vec{P}_{Kerr}(\vec{r},t)$. The discretization of this equation by employing the operators results to

$$\bar{\partial}_t\vec{D}^s = \left(\epsilon_0(\epsilon_\infty + \alpha\chi_0^{(3)}|\vec{E}^{s-\frac{1}{2}}|^2)\right)\bar{\partial}_t\vec{E}^s.$$

It is evident that $|\vec{E}^{s-\frac{1}{2}}|^2 = \widetilde{\vec{E}}_0\widetilde{\vec{E}}_0^{*}y^{2s-1}$ and this result leads to the characteristic polynomial

$$\Phi(y) = (y-1)^2(\epsilon_0(\epsilon_\infty + \alpha\chi_0^{(3)}|\widetilde{\vec{E}}_0|^2y^{2s-1})) + 4y\epsilon_0\epsilon_\infty\nu^2$$

Comparing this characteristic polynomial with this one obtained from the wave difference differential equation $\Phi'(y) = (y-1)^2\epsilon_0\epsilon_\infty + 4y\epsilon_0\epsilon_\infty\nu^2$, we see that they are similar when $\epsilon_\infty \gg \alpha\chi_0^{(3)}|\widetilde{\vec{E}}_0|^2y^{2s-1}$. Obviously, we assume that $|y| < 1$, but notice here that we are not assuming the same thing that we want to demonstrate. We know that $\Phi'(y)$, under the classical stability condition, leads to $|y| < 1$ and in general for $\forall s \geq \frac{1}{2}$ we find that $\Phi(y) \equiv \Phi'(y)$ when $\epsilon_\infty \gg \alpha\chi_0^{(3)}|\widetilde{\vec{E}}_0|^2$. We can identify the problem at a glance. The stability depends on the laser intensity $|\widetilde{\vec{E}}_0|$ through its propagation. In that case, we find the classical condition for stability already presented. Therefore, we can impose a condition to the laser intensity in order to preserve the stability; for instance, we can define an arbitrary factor D_f such that $\epsilon_\infty \simeq D_f\alpha\chi_0^{(3)}|\widetilde{\vec{E}}_0|^2$. Therefore, to ensure the stability $|\widetilde{\vec{E}}_0| < \sqrt{\frac{\epsilon_\infty}{D_f\alpha\chi_0^{(3)}}}$. From the last expression, we see that stability is related to the medium, optical properties as well as laser intensity. Hence a heuristic search helps to establish an asymptotic limit for this factor $D_f \geq 10^4$.

Appendix A.3. Stability Condition on Raman Effect

Again, we consider the wave equation plus the below difference equation for the Raman effect,

$$(1-\alpha)\epsilon_0\chi_0^{(3)}\frac{\tau_1^2+\tau_2^2}{\tau_1\tau_2^2}\left(\frac{\partial}{\partial t}\vec{E}(\vec{r},t)\int_0^t \bar{\bar{g}}(t-\tau)|\vec{E}(\vec{r},\tau)|^2 d\tau \right.$$
$$\left. + \bar{\bar{g}}(t)|\vec{E}(\vec{r},t)|^2\vec{E}(\vec{r},t)\right) + \epsilon_\infty\epsilon_0\frac{\partial}{\partial t}\vec{E}(\vec{r},t) = \frac{\partial}{\partial t}\vec{D}(\vec{r},t)$$

By applying the operators, we determine the discretized expression:

$$\epsilon_0\bar{\partial}_t\vec{E}^s\left(\epsilon_\infty + \Psi\int_0^{s\Delta_t}\bar{\bar{g}}(s\Delta_t - \tau)|\vec{E}(\vec{r},\tau)|^2 d\tau\right)$$
$$+\epsilon_0(1-\alpha)\chi_0^{(3)}\frac{\tau_1^2+\tau_2^2}{\tau_1\tau_2^2}\bar{\bar{g}}(s\Delta_t)\bar{\Delta}\left[\vec{E}^s|\vec{E}^s|^2\right] = \bar{\partial}_t\vec{D}^s$$

We calculate the characteristic polynomial replacing equations Equations (A1) and (A2) in the discretized wave equation and in the above expression. Then, we combine both derivation in the same formula,

$$\Phi(y) = (y-1)^2\Bigg(\epsilon_\infty + y^{2s-1}\Psi\Bigg(y\int_0^{s\Delta_t}\overline{\overline{g}}(s\Delta_t-\tau)d\tau + \frac{\Delta_t\overline{\overline{g}}(s\Delta_t)}{2}(y^2+y+1)\Bigg)|\tilde{\vec{E}}_0|^2\Bigg) + 4y\epsilon_\infty\nu^2$$

Under an approach analogous to the one utilizes in the previous case we have to link the power laser to the medium. That means that the scheme should verify the inequality,

$$\epsilon_\infty \gg y^{2s-1}\Psi\Bigg(y\int_0^{s\Delta_t}\overline{\overline{g}}(s\Delta_t-\tau)d\tau + \frac{\Delta_t\overline{\overline{g}}(s\Delta_t)}{2}(y^2+y+1)\Bigg)|\tilde{\vec{E}}_0|^2.$$

We know that $\int_0^{s\Delta_t}\overline{\overline{g}}(s\Delta_t-\tau)d\tau \gg \frac{\Delta_t\overline{\overline{g}}(s\Delta_t)}{2}$, hence, a reasoning similar to the previous section applies and we establish the stability condition $|\tilde{\vec{E}}_0| < \sqrt{\frac{\epsilon_\infty}{D_f(1-\alpha)\chi_0^{(3)}\frac{\tau_1^2+\tau_2^2}{\tau_1\tau_2^2}\int_0^{s\Delta_t}\overline{\overline{g}}(s\Delta_t-\tau)d\tau}}$. Finally for a reason of normalization we find that $|\tilde{\vec{E}}_0| < \sqrt{\frac{\epsilon_\infty}{D_f(1-\alpha)\chi_0^{(3)}}}$, together with the classic condition $\Delta_t \leq \frac{n_i}{c_0}\sqrt{\frac{1}{\Delta_x^2}+\frac{1}{\Delta_y^2}+\frac{1}{\Delta_z^2}}$ are the stability conditions in this case. Once again, we observe that stability is related to the medium optical properties as well as the laser intensity when the Raman nonlinear effect is present.

Appendix A.4. Stability Condition on Photoionization Effect

Together the wave equation we take into consideration the differential equation $\frac{\partial}{\partial t}\vec{D}(\vec{r},t) = \epsilon_\infty\epsilon_0\frac{\partial}{\partial t}\vec{E}(\vec{r},t) + \sigma_{fe^-}(\vec{r},t)\vec{E}(\vec{r},t)$, which by means of the defined operators can be discretized as follows,

$$\bar{\partial}_t\vec{D}^s = \epsilon_\infty\epsilon_0\bar{\partial}_t\vec{E}^s + \sigma_{fe^-}\bar{\Delta}\vec{E}^s$$

Again we combine the equation with the discretized wave differential equation with the aim to determine the characteristic polynomial $\Phi(y) = 2\epsilon_0\epsilon_\infty(1+y(4\nu^2-2)+y^2) + \Delta_t\sigma_{fe^-}(y^2-1)$ in which we apply the bilinear transformation $y = \frac{r+1}{r-1}$. Therefore, we calculate the characteristic polynomial $\Phi(r) = 2\epsilon_0\epsilon_\infty\nu^2r^2 + \sigma_{fe^-}\Delta_t r + 2\epsilon_0\epsilon_\infty(1-\nu^2)$ for which the Routh-Hurwitz criterion can be applied. The first column of the Routh table are formed by the coefficients $c_{1,0} = 2\epsilon_0\epsilon_\infty\nu^2$, $c_{2,0} = \sigma_{fe^-}\Delta_t$ and $c_{3,0} = 2\epsilon_0\epsilon_\infty(1-\nu^2)$. The later coefficient provides the classical condition for stability. In addition, we know that σ_{fe^-} must be a positive magnitude so as to address the stability in the algorithm updating.

Appendix B. Computational Resources

Table A1 lists the computational resources employed in the simulations. The operative system in which the simulations were carried out and runs on this hardware is Ubuntu Bionic Beaver.

Table A1. Computational resources.

Computational Resources:
CPU: Intel Core i9-10980XE
MB: GIGABYTE X299X
RAM: Crucial CT32G4RF D4293 DDR4-2933
ROM: 2x Crucial P5 1TB PCIe M.2 2280SS SSD
GPU: GeForce GTX 1080 Ti

References

1. Greene, J.H.; Taflove, A. General vector auxiliary differential equation finite-difference time-domain method for nonlinear optics. *Opt. Express* **2006**, *14*, 8305–8310, doi:10.1364/OE.14.008305.
2. Taflove, A.; Oskooi, A.; Johnson, S. *Advances in FDTD Computational Electrodynamics: Photonics and Nanotechnology*; Artech House Antennas and Propagation Library; Artech House: Norwood, MA, USA, 2013.
3. Maksymov, I.S.; Sukhorukov, A.A.; Lavrinenko, A.V.; Kivshar, Y.S. Comparative Study of FDTD-Adopted Numerical Algorithms for Kerr Nonlinearities. *IEEE Antennas Wirel. Propag. Lett.* **2011**, *10*, 143–146, doi:10.1109/LAWP.2011.2114319.
4. Hellwarth, R. Third-order optical susceptibilities of liquids and solids. *Prog. Quantum Electron.* **1977**, *5*, 1–68, doi:10.1016/0079-6727(79)90002-8.
5. Jia, H.; Li, J.; Fang, Z.; Li, M. A new FDTD scheme for Maxwell's equations in Kerr-type nonlinear media. *Numer. Algorithms* **2019**, *82*, 223–243.
6. Weber, W.H.; Ford, G.W. Optical electric-field enhancement at a metal surface arising from surface-plasmon excitation. *Opt. Lett.* **1981**, *6*, 122–124, doi:10.1364/OL.6.000122.
7. Yu, H.; Peng, Y.; Yang, Y.; Li, Z.Y. Plasmon-enhanced light–matter interactions and applications. *NPJ Comput. Mater.* **2019**, *5*, 1–14.
8. Moreno, E.; Sohrabi, R.; Klochok, G.; Michael, E. Vertical versus planar pulsed photoconductive antennas that emit in the terahertz regime. *Optik* **2018**, *166*, 257–269.
9. Blow, K.J.; Wood, D. Theoretical description of transient stimulated Raman scattering in optical fibers. *IEEE J. Quantum Electron.* **1989**, *25*, 2665–2673, doi:10.1109/3.40655.
10. Duchateau, G.; Bourgeade, A. Influence of the time-dependent pulse spectrum on ionization and laser propagation in nonlinear optical materials. *Phys. Rev. A* **2014**, *89*, 053837.
11. Schmitz, H.; Mezentsev, V. Full-vectorial modeling of femtosecond pulses for laser inscription of photonic structures. *J. Opt. Soc. Am. B* **2012**, *29*, 1208–1217.
12. Bulgakova, N.M.; Zhukov, V.P.; Sonina, S.V.; Meshcheryakov, Y.P. Modification of transparent materials with ultrashort laser pulses: What is energetically and mechanically meaningful? *J. Appl. Phys.* **2015**, *118*, 233108.
13. Ardaneh, K.; Giust, R.; Morel, B.; Courvoisier, F. Generation of a Bessel beam in FDTD using a cylindrical antenna. *Opt. Express* **2020**, *28*, 2895–2908.
14. Nguyen, H.D.; Sedao, X.; Mauclair, C.; Bidron, G.; Faure, N.; Moreno, E.; Colombier, J.P.; Stoian, R. Non-Diffractive Bessel Beams for Ultrafast Laser Scanning Platform and Proof-Of-Concept Side-Wall Polishing of Additively Manufactured Parts. *Micromachines* **2020**, *11*, 974.
15. Bévillon, E.; Stoian, R.; Colombier, J.P. Nonequilibrium optical properties of transition metals upon ultrafast electron heating. *J. Phys. Condens. Matter* **2018**, *30*, 385401.
16. Bourgeade, A.; Duchateau, G. Time-dependent ionization models designed for intense and short laser pulse propagation in dielectric materials. *Phys. Rev. E* **2012**, *85*, 056403.
17. Sommer, A.; Bothschafter, E.M.; Sato, S.A.; Jakubeit, C.; Latka, T.; Razskazovskaya, O.; Fattahi, H.; Jobst, M.; Schweinberger, W.; Shirvanyan, V.; et al. Attosecond nonlinear polarization and light–matter energy transfer in solids. *Nature* **2016**, *534*, 86–90.
18. Couairon, A.; Sudrie, L.; Franco, M.; Prade, B.; Mysyrowicz, A. Filamentation and damage in fused silica induced by tightly focused femtosecond laser pulses. *Phys. Rev. B* **2005**, *71*, 125435.
19. Holzmann, G.J. *The Power of Ten-Rules for Developing Safety Critical Code*; Technical Report; NASA/JPL Laboratory for Reliable Software: Pasadena, CA, USA, 2020.
20. Lax, P. *Hyperbolic Systems of Conservation Laws and the Mathematical Theory of Shock Waves*; CBMS-NSF Regional Conference Series in Applied Mathematics; Society for Industrial and Applied Mathematics: Philadelphia, PA, USA, 1973.
21. Wu, A.Q.; Chowdhury, I.H.; Xu, X. Femtosecond laser absorption in fused silica: Numerical and experimental investigation. *Phys. Rev. B* **2005**, *72*, 085128.
22. Taflove, A.; Hagness, S.C. *Computational Electrodynamics the Finite-Differences Time Domain Method*; Artech House Publishers: Norwood, MA, USA, 2005.
23. Potter, M.; Bérenger, J.P. A Review of the Total Field/Scattered Field Technique for the FDTD Method. *FERMAT* **2017**, *19*, 1–13.
24. Moreno, E.; Hemmat, Z.; Roldán, J.B.; Pantoja, M.F.; Bretones, A.R.; García, S.G.; Faez, R. Implementation of Open Boundary Problems in Photo-Conductive Antennas by Using Convolutional Perfectly Matched Layers. *IEEE Trans. Antennas Propag.* **2016**, *64*, 4919–4922.
25. Tan, T.; Potter, M. On the Nature of Numerical Plane Waves in FDTD. *IEEE Antennas Wirel. Propag. Lett.* **2009**, *8*, 505–508.
26. Tan, T.; Potter, M. FDTD Discrete Planewave (FDTD-DPW) Formulation for a Perfectly Matched Source in TFSF Simulations. *IEEE Trans. Antennas Propag.* **2010**, *58*, 2641–2648.
27. Sohrabi, R.; Moreno, E.; Michael, E.A. Total field scattering field formulation in a non-uniform grid by means of FDTD-DPW. *Electromagnetics* **2020**, *40*, 40–55, doi:10.1080/02726343.2019.1695089.
28. Rudenko, A. Étude NuméRique de la Formation des Nanostructures Périodiques Induites par Laser Ultrabref Dans les MatéRiaux Diélectriques. Ph.D. Thesis, Université Jean-Monnet, Saint-Étienne, France, 2017.
29. Gulley, J.R. Frequency dependence in the initiation of ultrafast laser-induced damage. In Proceedings of the Laser Damage Symposium XLII: Annual Symposium on Optical Materials for High Power Lasers, Boulder, CO, USA, 26–29 September 2010.
30. Malitson, I.H. Interspecimen Comparison of the Refractive Index of Fused Silica. *J. Opt. Soc. Am.* **1965**, *55*, 1205–1209.

31. Stratton, J.A. *Electromagnetic Theory*; Wiley: Hoboken, NJ, USA, 2015.
32. Yang, W.; Bricchi, E.; Kazansky, P.G.; Bovatsek, J.; Arai, A.Y. Self-assembled periodic sub-wavelength structures by femtosecond laser direct writing. *Opt. Express* **2006**, *14*, 10117–10124.
33. Zhang, F.; Nie, Z.; Huang, H.; Ma, L.; Tang, H.; Hao, M.; Qiu, J. Self-assembled three-dimensional periodic micro-nano structures in bulk quartz crystal induced by femtosecond laser pulses. *Opt. Express* **2019**, *27*, 6442–6450.
34. Bhardwaj, V.; Simova, E.; Rajeev, P.; Hnatovsky, C.; Taylor, R.; Rayner, D.; Corkum, P. Optically produced arrays of planar nanostructures inside fused silica. *Phys. Rev. Lett.* **2006**, *96*, 057404.
35. Rudenko, A.; Colombier, J.P.; Itina, T.E. Nanopore-mediated ultrashort laser-induced formation and erasure of volume nanogratings in glass. *Phys. Chem. Chem. Phys.* **2018**, *20*, 5887–5899.
36. Sugioka, K. Progress in ultrafast laser processing and future prospects. *Nanophotonics* **2017**, *6*, 393–413.
37. Watanabe, W.; Asano, T.; Yamada, K.; Itoh, K.; Nishii, J. Wavelength division with three-dimensional couplers fabricated by filamentation of femtosecond laser pulses. *Opt. Lett.* **2003**, *28*, 2491–2493.
38. Florea, C.; Winick, K.A. Fabrication and characterization of photonic devices directly written in glass using femtosecond laser pulses. *J. Light. Technol.* **2003**, *21*, 246.
39. Bricchi, E.; Mills, J.D.; Kazansky, P.G. Klappauf, B.G.; Baumberg, J.J. Birefringent Fresnel zone plates in silica fabricated by femtosecond laser machining. *Opt. Lett.* **2002**, *27*, 2200–2202.
40. Pereda, A.; Vielva, L.A.; Vegas, A.; Prieto, A. Analyzing the stability of the FDTD technique by combining the von Neumann method with the Routh-Hurwitz criterion. *IEEE Trans. Microw. Theory Tech.* **2001**, *49*, 377–381, doi:10.1109/22.903100.
41. Moreno, E.; Cruz-Hernández, P.; Hemmat, Z.; Mojab, A.; Michael, E. Comparative study of grids based on the cubic crystal system for the FDTD solution of the wave equation. *Math. Comput. Simul.* **2019**, *166*, 395–409.
42. Hui, F.; Liu, P.; Hodge, S.A.; Carey, T.; Wen, C.; Torrisi, F.; Galhena, D.T.L.; Tomarchio, F.; Lin, Y.; Moreno, E.; et al. In-situ Observation of Low-Power Nano-Synaptic Response in Graphene Oxide using Conductive Atomic Force Microscopy. *Nano Micro Small* **2021**, *17*, 2101100.

applied sciences

MDPI

Article

Incident Angle Dependence of the Waveform of the Polarization-Sensitive Photoresponse in CuSe/Se Thin Film

Arseniy E. Fateev, Tatyana N. Mogileva, Vladimir Ya. Kogai, Konstantin G. Mikheev and Gennady M. Mikheev *

Institute of Mechanics, Udmurt Federal Research Center, Ural Branch of the Russian Academy of Sciences, Tatyana Baramzinoy Street, 34, 426067 Izhevsk, Russia; a.e.fateev@mail.ru (A.E.F.); mogileva@udman.ru (T.N.M.); vkogai@udman.ru (V.Y.K.); k.g.mikheev@udman.ru (K.G.M.)
* Correspondence: mikheev@udman.ru; Tel.: +7-3412-21-66-11

Abstract: The results of studying the waveforms of longitudinal and transverse photocurrent pulses generated in thin, semitransparent CuSe/Se films as a function of the angle of incidence (α) of a femtosecond laser beam at linear and circular polarizations are presented. It has been established that the durations of unipolar longitudinal photocurrent pulses at linear and circular polarizations of laser pumping do not depend on the angle α. It is shown that the evolution of the temporal profile of the helicity-sensitive transverse photocurrent with a change in α strongly depends on polarization. At linear polarization, the shape of the unipolar pulses remains virtually constant; however, at circular polarization, the generation of unipolar and bipolar pulses is possible, with the waveforms strongly depending on the angle α. The influence of the incidence angle on the waveforms of transverse photocurrent pulses is explained by the transformation of linear and circular polarization into an elliptical upon the refraction of light at the air/semitransparent film interface and by the interplay of photocurrents arising due to linear and circular surface photogalvanic effects in the film. The presented findings can be utilized to develop polarization and incidence angle-sensitive photovoltaic devices.

Keywords: circular photocurrent; polarization; helicity; bipolar photovoltage; waveforms; femtosecond excitation; surface photogalvanic effect; thin films

Citation: Fateev, A.E.; Mogileva, T.N.; Kogai, V.Y.; Mikheev, K.G.; Mikheev, G.M. Incident Angle Dependence of the Waveform of the Polarization-Sensitive Photoresponse in CuSe/Se Thin Film. *Appl. Sci.* **2022**, *12*, 6869. https://doi.org/10.3390/app12146869

Academic Editor: Alejandro Pérez-Rodróhiez

Received: 12 May 2022
Accepted: 5 July 2022
Published: 7 July 2022

1. Introduction

One of the interesting features of the interaction of polarized light [1–3] with semiconductors and metals is the generation of a polarization-sensitive photocurrent (PSPC) that depends on the polarization of the incident radiation according to harmonic laws [4–9]. There are longitudinal and transverse PSPCs at the oblique incidence of light on the surface of a material in which the currents flow along and perpendicular to the plane of incidence, respectively [10]. The transverse PSPC can be a combination of circular photocurrents (CPC) and linear photocurrents (LPC). The CPC depends on the direction of rotation of the electric field vector (the sign of circular polarization) of the incident radiation, while the LPC does not [11–14]. It is of interest to study the laws governing the PSPC in various media in terms of excitation of spin-polarized electrons [15,16], for the creation and development of optospintronic devices [17], laser polarization analyzers [18–20], photodiodes with high spatial resolution [21], as well as for photosensors intended for the direct recording of the polarization state of circularly polarized light [22–25]. The mechanisms of the PSPC generation include the photogalvanic effect (PGE) [26–28], the circular photogalvanic effect (CPGE) [11,29–34], the photon drag effect (PDE) [5,10,35–38], and the surface photogalvanic effect (SPGE) [5,39,40]. All of the aforementioned effects are nonlinear optical phenomena. It should be noted that the PDE and SPGE, as well as some other nonlinear optical phenomena [41,42] can be observed in any media, regardless of the type of symmetry of the medium [35]. The PDE and SPGE photocurrent pulses excited by pulsed laser pumping are usually unipolar (see, for example, [14,43–46]). The dependence of CPC and LPC excited in film and 2D structures due to the PDE and SPGE on the beam incidence angle α is described by an odd function [36,37,39,46–48]. Therefore, under pulsed laser pumping, the LPC and CPC unipolar pulses change their polarity when the sign of the incidence angle changes. With the simultaneous occurrence of longitudinal PDE and SPGE, due to their interplay, it is possible to generate bipolar PSPC pulses consisting of front and tail parts of opposite polarity, which change their polarities when the sign of the angle of incidence changes [40]. It was shown in our most recent work [48] that, at a given angle of incidence, the interference of the CPC and LPC generated due to the SPGE in a CuSe/Se nanocomposite film also leads to the generation of bipolar photocurrent pulses. The temporal profiles of those pulses strongly depend on the polarization ellipse of the incident femtosecond laser beam. However, despite the large number of articles on the topic of PSPC generation in various film materials (see, for example, [33,38,49–52]), studies of the influence of the incidence angle on the waveforms of photocurrent pulses arising due to the PDE and SPGE (and also CPGE) under pulsed laser pumping have not been carried out yet.

In this work, using the study of the SPGE in a CuSe/Se semitransparent thin film under femtosecond laser excitation as an example, it is shown for the first time that the waveforms of the PSPC pulses can strongly depend on the angle of incidence. In particular, it has been shown that with circularly polarized pulsed pumping and a fixed sign of the incidence angle, unipolar transverse photocurrent pulses of opposite polarity are generated at small and large angles of incidence, and bipolar pulses are excited in the intermediate range of incident angles, smoothly transforming into unipolar pulses of opposite polarity at the boundaries of this interval. The results obtained are explained by the change in the state of polarization in the refracted beam and by the interference of the LPC and CPC that arises in the subsurface layer of the semiconducting medium under study.

2. Materials and Methods

Thin CuSe/Se films with a thickness of 130 nm and dimensions of 15 × 35 mm were synthesized on a glass substrate by successive vacuum thermal deposition of Se and Cu, according to the procedure described in one of our previous publications [14]. To stabilize the residual Se [53,54], the synthesized film structure was annealed at a temperature of 140 °C for 30 min. As a result, the selenium from an unstable amorphous phase was transformed into a stable polycrystalline trigonal selenium (*t*-Se). It should be noted that currently there are some other methods of film copper selenides structure synthesis that have also been developed [55–60].

The phase composition of the films was studied at room temperature on a Bruker D2 PHASER diffractometer. We used CuKα radiation (λ = 0.154 nm). Using the DIFFRAC.EVA universal program, the diffraction curves were smoothed and the background due to X-ray scattering on a glass substrate was subtracted. To study the phase composition of the films, we additionally used a HORIBA HR800 Raman spectrometer with laser excitation at 632.8 nm at the radiation intensity of 14 kW/cm^2. Sample surface images were taken using a scanning electron microscope (SEM) (Thermo Fisher Scientific Quattro S, Brno-Černovice, Czech Republic). The elemental composition of the synthesized films was determined using an energy-dispersive microanalyzer based on an EDAX "Octane Elect Plus EDS System" spectrometer built into the SEM. The optical transmittance spectra of the films were recorded using a two-beam spectrophotometer (PerkinElmer Lambda 650, Shelton, WA, USA). The film thickness was measured using a stylus profilometer. The refractive index n and the extinction κ of the film were determined using an ellipsometer.

The X-ray diffraction pattern (see Figure S1, Supporting Information) and the Raman spectrum (see Figure S2, Supporting Information) demonstrate that the film has two phases and consists of CuSe and *t*-Se. The mass percentages of CuSe and *t*-Se in the synthesized film are 33.3 and 66.7 wt.%, respectively. The film surface consists of a number of flat petal-like structures, which are CuSe crystallites predominantly oriented in radial directions from their centers (see Figure S3, Supporting Information). The centers of these structures are located at distances of about 15–30 μm from each other. The film is semitransparent in the visible wavelength range (see Figure S4, Supporting Information). The optical transmittance, refractive index n, and extinction coefficient κ at 795 nm are 27.6%, 1.64, and 2.14, respectively. The synthesized film is electrically conductive with a sheet resistance of 39 $\Omega/\square$, which is orders of magnitude lower than the corresponding value of many known thin-film semiconductor structures (see, for example, [61,62]). It should be added that the sheet resistance of CuSe thin films with thicknesses from 50 to 500 nm synthesized by the chemical bath method varies between 23–50 $\Omega/\square$ [63]. Meanwhile, the sheet resistance of 280 nm thick bismuth selenide (Bi_2Se_3) film obtained by the same method immediately after synthesis is 10^{12} $\Omega/\square$, and after its annealing in air or in the nitrogen medium, the sheet resistance decreases to 3.6×10^3 $\Omega/\square$ [64].

To measure the laser-induced photocurrent (photovoltage), two film gold electrodes A and B were deposited on opposite short sides of the synthesized film by vacuum thermal deposition. The film under study was placed on a special goniometric device, which made it possible to smoothly change its spatial orientation relative to the incident laser beam. The surface of the film remaining after depositing the electrodes was sufficient to study the generation of photocurrent under the action of a narrow laser beam 1.5 mm in diameter in a wide range of incidence angles.

The photocurrent pulses in a CuSe/*t*-Se nanocomposite film were excited by pulses of a Ti:S femtosecond laser at a wavelength of 795 nm (pulse duration 120 fs, pulse repetition rate 1 kHz) at an oblique incidence of radiation on the film. All the experiments were carried out at laser pulse energy E_{in} = 100 μJ. The polarization of the radiation incident on the film was controlled using half-wave and quarter-wave plates. The photocurrent pulses were registered and recorded using a digital oscilloscope with a bandwidth of 400 MHz and a rise time of 875 ps with averaging over multiple exciting laser pulses. The input impedance of the oscilloscope was R_{in} = 50 Ω. In the experiments, the extreme values of the pulses of the longitudinal and transverse photocurrents j_x and j_y, respectively, were recorded, determined by the formulas $j_x = U_x/R_{in}$ and $j_y = U_y/R_{in}$, where U_x and U_y are the extreme values of the voltage pulses at the experimental geometries when the measuring electrodes were placed perpendicular and parallel to the plane of incidence, respectively. Since U_x and U_y depend linearly on E_{in} [65], it was convenient to introduce the conversion coefficients $\eta_x = j_x/E_{in}$ and $\eta_y = j_y/E_{in}$, which characterize the efficiency of converting light into longitudinal and transverse photocurrents, respectively. It should be noted that when studying the waveforms of the photocurrent pulses, we subtracted the time-varying background signal due to noticeable electrical interference coming from the power supply unit of the femtosecond laser.

3. Results and Discussion

Figure 1a shows the dependences of the coefficients of light conversion into longitudinal photocurrent η_x on the angle of incidence α for linear and circular polarizations, measured according to the sketch of experimental arrangements presented in the upper frame. It can be seen that the photocurrent is absent at normal incidence of the radiation on the film (α = 0). The photocurrent changes polarity when the sign of the angle of incidence changes, increases in absolute value with an increase in the modulus of α, takes its extreme values at angles $|\alpha| \approx 66$–$68°$, and disappears at a grazing incidence of radiation on the film. Thus, the longitudinal photocurrent for linear and circular polarizations is described by an odd function of the angle α:

$$\eta_x(\alpha) = -\eta_x(-\alpha). \quad (1)$$

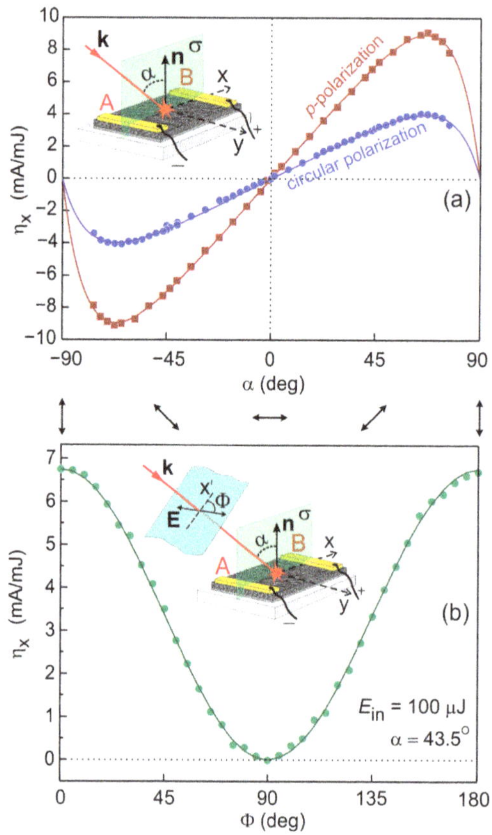

Figure 1. The light to longitudinal photocurrent conversion coefficient η_x as a function of (**a**) the light incidence angle α on the film at linear (p-polarization) and circular polarizations (points denote experimental results, curves denote fitting with $\eta_x = 35.37 \sin 2\alpha/(1.76 \cos\alpha + 1)^2$ and $\eta_x = 15.68 \sin 2\alpha/(1.75 \cos\alpha + 1)^2$ for linear and circular polarization, respectively) and (**b**) the azimuth of the polarization plane Φ at α = 43.5° (dots denote experimental results, curve denotes fitting with $\eta_x(\Phi) = 6.7 \cos^2\Phi$; orientations of the polarization plane are shown at the top). The insets show the sketches of the experimental setups: **k** and **E** are the wave vector and the electric field vector of the incident radiation, respectively; σ is the plane of incidence; α is an angle of incidence; **n** is the normal to the film surface; **A** and **B** are the measuring electrodes; x, y are the axes of the rectangular coordinate system; the x and x′ axes lie in the σ plane (x′ $\perp$ **k**, the x axis is perpendicular to the electrodes **A** and **B**); Φ is the angle between the x′ and **E**.

With linear polarization, the dependence of η_x on the polarization azimuth Φ (the angle between the plane of polarization and the radiation incidence plane on the film σ) at a fixed α is described by an even function:

$$\eta_x(\alpha, \Phi) = \eta_x(\alpha, \Phi = 0) \cos^2, \tag{2}$$

where $\eta_x(\alpha, \Phi = 0)$ is the conversion coefficient at a given α and Φ = 0. It should be noted that Φ = 0 corresponds to p-polarized radiation. All this is clearly seen from the dependence $\eta_x(\Phi) = \eta_x(\alpha, \Phi = 0) \cos^2\Phi$ obtained at α = 43.5°, where $\eta_x(\alpha, \Phi = 0) = 6.7$ mA/mJ (see Figure 1b). From Equation (2) it follows that $\eta_x(\alpha = 43.5°, \Phi = 90°) = 0$, i.e., no photocurrent is generated at s-polarization. Experiments have shown that this equation is valid for any α. Figure 2a shows the dependence of the conversion coefficient of light into transverse photocurrent η_y on the angle of incidence α for linearly polarized radiation at polarization azimuth $\Phi = -45°$, obtained using the experimental setup shown in the upper inset to this figure. It can be seen that, similarly to the angular dependence of the longitudinal photocurrent, the experimental dependence $\eta_y(\alpha)$ is described by an odd function, i.e.:

$$\eta_y(\alpha) = -\,\eta_y(-\alpha). \tag{3}$$

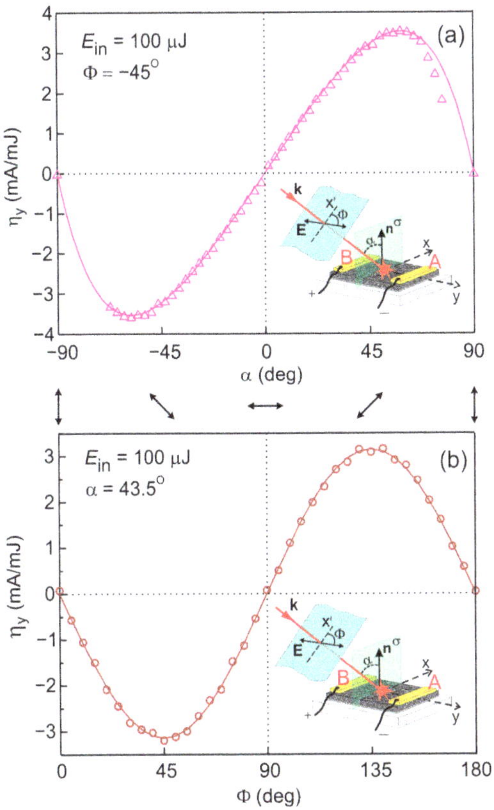

Figure 2. The light to transverse photocurrent conversion coefficient η_y as a function of (**a**) the incidence angle α at the polarization azimuth $\Phi = -45°$ (triangles correspond to the experimental data while the solid line represents the result of the fitting with equation $\eta_y = 8.37 \sin 2\alpha/(0.87\cos\alpha + 1)^2$) and (**b**) the polarization azimuth Φ at $\alpha = 43.5°$ (circles denote experimental data, the solid curve represents the fitting with the equation $\eta_y = -3.1 \sin 2\,\Phi$; the orientations of the plane of polarization of the incident radiation are shown at the top). The insets show the sketches of the experimental setups.

Meanwhile, the measured dependence of the transverse photocurrent on the polarization azimuth Φ at a fixed $\alpha = 43.5°$ (see Figure 2b) is approximated by an odd function:

$$\eta_y(\alpha = 43.5°, \Phi) = \eta_y(\alpha = 43.5°, \Phi = 45°) \sin 2\Phi, \tag{4}$$

where $\eta_y(\alpha = 43.5°, \Phi = 45°)$ is the conversion coefficient of light into transverse photocurrent at $\alpha = 43.5°$ and $\Phi = 45°$.

The set of angular and polarization dependences described by Equations (1)–(4) shown in Figures 1 and 2 indicates that the generation of photocurrent in the studied CuSe/*t*-Se thin films occurs according to the SPGE mechanism [5,39,40,65].

Figure 3a shows the unipolar pulse waveforms of the longitudinal photovoltage (measured in the geometry of the experiment when the measuring electrodes are oriented perpendicular to the plane of incidence), normalized to their extreme values at *p*-polarization of the incident radiation at $\alpha = \pm 45^\circ$ and $\alpha = \pm 78^\circ$. It follows that when the sign of the incidence angle is changed, the pulses are inverted, which is in agreement with Equation (1). It can also be seen that the waveforms of the pulses are virtually independent of the angle of incidence. This is evidenced by the dependence curve of the pulse duration τ (FWHM) on α, presented in the inset to Figure 3a. Similar patterns were obtained when the longitudinal photovoltage was excited by circularly polarized radiation pulses (see Figure 3b). Thus, the waveforms of the longitudinal photovoltage pulses in the films under study for linear and circular polarizations are virtually independent from the angle of incidence.

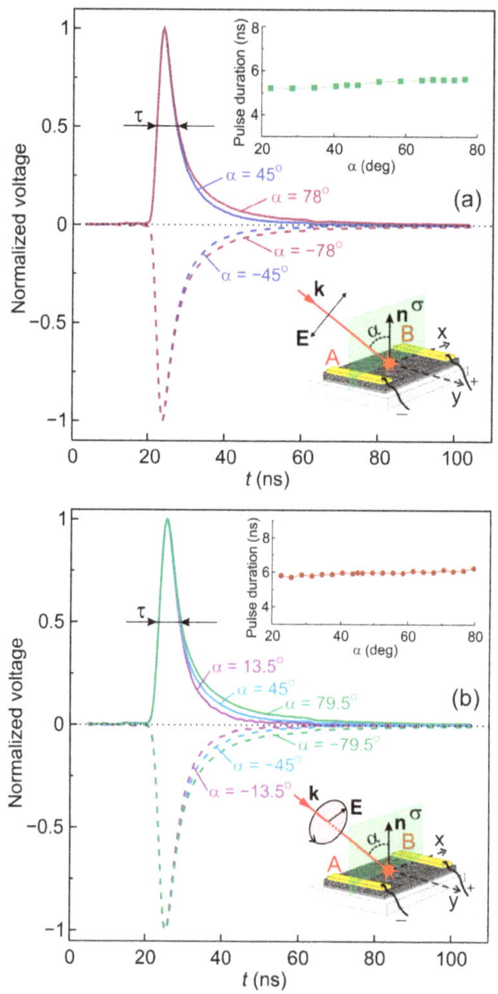

Figure 3. The waveforms of the longitudinal photovoltage pulses normalized to their extreme values at (**a**) linear polarization ($\Phi = 0$, *p*-polarization) and (**b**) circular polarization recorded at positive (solid lines) and negative (dashed lines) incidence angles α. The upper insets show the corresponding dependences of the recorded pulse durations τ on the angle of incidence. The bottom insets show the sketches of the experimental setups.

The waveforms of the transverse photovoltage pulses recorded at linear polarization ($\Phi = -45°$) and normalized to their extreme values at different α are shown in Figure 4. It can be seen that the waveforms of these pulses weakly depend on α. This is evidenced by the dependence of the pulse duration on the incidence angle, shown in the inset in the same figure.

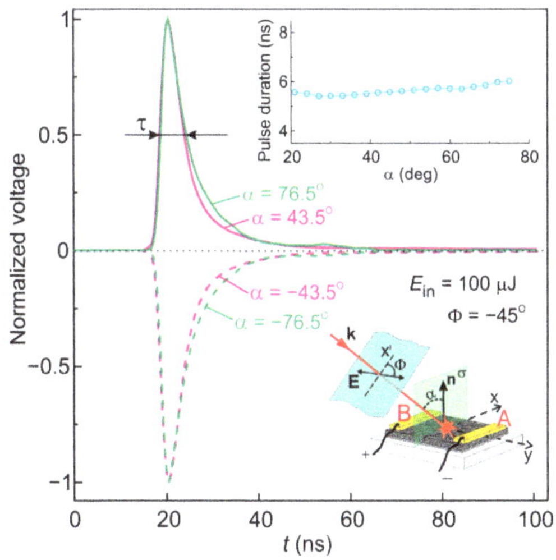

Figure 4. The waveforms of the transverse photovoltage pulses normalized to their extreme values at linear polarization ($\Phi = -45°$), recorded at positive (solid lines) and negative (dashed lines) angles of incidence α. The upper inset shows the dependence of the recorded pulses duration τ on the angle of incidence. The bottom inset shows the sketch of the experimental setup.

Figure 5 shows the waveforms of the transverse photovoltage pulses recorded for the right-hand circularly polarized laser beam at various angles of incidence. The pulses are normalized to their extreme values. It can be seen that at $0 < \alpha < 58.5°$, positive pulses are generated, the duration of which gradually increases with increasing α. In the range of angles α approximately $58.5 \leq \alpha \leq 76.5°$ bipolar pulses are excited. Figure 5 shows that at $\alpha = 58.5°$, a small negative pulse appears on the leading edge of the positive pulse. The amplitude of the front negative pulse increases with increasing the angle of incidence, while the amplitude of the positive tail of the pulse decreases. At large angles of incidence ($\alpha = 79.5°$ and $82.5°$), the positive part of the pulse completely disappears and the photovoltage is generated in the form of a negative unipolar pulse. It should be added that when the direction of rotation of the electric field vector of the incident radiation changes (i.e., at left-hand circularly polarized laser beam) the polarities of the pulses shown in Figure 5 change to the opposite ones. Thus, the waveforms of the transverse photovoltage pulses at circular polarization strongly depend on the angle of incidence, which is not typical of the longitudinal photovoltage pulses at different polarizations (see Figure 3) and the transverse photovoltage pulses at linear polarization (see Figure 4).

The unusual dependence of the waveform of the transverse photovoltage pulses on the angle of incidence at circular polarization can be explained on the basis of the interaction of linear and circular photocurrents arising in the surface layer of the film as a result of the transformation of the incident circular polarization into elliptical polarization upon the refraction of light at the air/film interface.

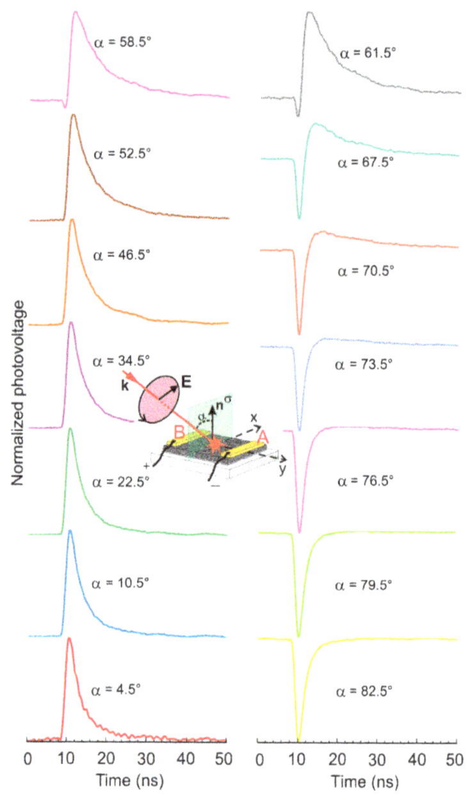

Figure 5. The waveforms of the transverse photovoltage pulses normalized to their extreme values for circularly polarized incident radiation, recorded for different angles of incidence α. The inset shows the sketch of the experimental setup.

To determine the polarization of the refracted beam, one can use the complex Fresnel refractive indices t_p, t_s for p- and s-polarizations, respectively, at the air/semitransparent film interface given in [66]:

$$t_p = \frac{2\hat{n}\cos\alpha}{\hat{n}^2\cos\alpha + \sqrt{\hat{n}^2 - \sin^2\alpha}}, t_s = \frac{2\cos\alpha}{\sqrt{\hat{n}^2 - \sin^2\alpha} + \cos\alpha}, \quad (5)$$

where $\hat{n} = n + i\kappa$ is the complex refractive index, n is the real refractive index that determines the phase velocity (n = 1.64), and κ is the absorption coefficient (κ = 2.14) that determines the attenuation of light in the film itself. From expression (5), the phase shift δ_t between the components of the refracted beam with p- and s-polarizations is determined by the formula:

$$\delta_t = -\delta_{tp} + \delta_{ts} + \delta_0, \quad (6)$$

where:

$$\delta_{tp} = \text{Arg}(t_p) = \tan^{-1}\left(\frac{-\kappa^2\cos\alpha + \chi\cos\xi + n^2\cos\alpha}{\chi\sin\xi + 2\kappa n\cos\alpha}\right) - \tan^{-1}\left(\frac{n}{\kappa}\right), \quad (7)$$

$$\delta_{ts} = \text{Arg}(t_s) = \tan^{-1}\left(\frac{(\cos\alpha + \chi\cos\xi)\csc\xi}{\chi}\right) - \frac{\pi}{2}, \quad (8)$$

$$\chi = \sqrt[4]{\left(-\sin^2\alpha - \kappa^2 + n^2\right)^2 + 4\kappa^2 n^2}, \tag{9}$$

$$\xi = \frac{1}{2\left(\frac{\pi}{2} - \tan^{-1}\left(\frac{-\sin^2\alpha - \kappa^2 + n^2}{2\kappa n}\right)\right)}, \tag{10}$$

δ_{tp}, δ_{ts} are the phase shifts of the p- and s-beam components, respectively, resulting from refraction at the interface between two media with a complex refractive index, and δ_0 is the initial phase shift between the p- and s-components before refraction. For the circular polarization of the incident radiation $\delta_0 = \pi/2$, where the signs "+" and "−" are the opposite of the sign of circular polarization.

It follows from Equation (5) that the amplitude values of the complex transmission coefficients t_p and t_s can be determined using:

$$|t_p| = \frac{2\cos\alpha\sqrt{\kappa^2 + n^2}}{\sqrt{\left(-\kappa^2\cos\alpha + \chi\cos\xi + n^2\cos\alpha\right)^2 + \left(2\kappa n\cos\alpha + \chi\sin\chi\right)^2}}, \tag{11}$$

$$|t_s| = \frac{2\cos\alpha}{\sqrt{\chi^2\sin^2\xi + \left(\cos\alpha + \chi\cos\xi\right)^2}}. \tag{12}$$

The amplitudes of the p- and s-components of the electric field vector of the refracted beam $\mathbf{E}^{(t)}$ describing the ellipse in the x″y coordinate system can be found using the equations:

$$T_{x''} = |t_p|E_p;\ T_y = |t_s|E_s, \tag{13}$$

respectively, where x″y is the coordinate plane perpendicular to the wave vector $\mathbf{k}_t$ of the refracted beam, axis x″ is in the plane of incidence σ, and the axis y is perpendicular to σ.

In this case, the equation of the polarization ellipse in the refracted beam depending on the ratio of the p- and s-components of the vector $\mathbf{E}^{(t)}$ and the phase shift δ_t can be written as follows:

$$\frac{(x'')^2}{T_{x''}{}^2} + \frac{y^2}{T_y^2} - \frac{2x''y\cos\delta_t}{T_{x''}T_y} = \sin^2\delta_t. \tag{14}$$

The polarization ellipses are characterized by the angle ψ between the semi-major axis a of the ellipse and the x″ axis lying in the incidence plane σ of radiation on the film (see Figure 6a,b right insets), as well as the degree of circular polarization $P_{cir} = \gamma 2ab/(a^2 + b^2)$ and degree of linear polarization $P_{lin} = (a^2 - b^2)/(a^2 + b^2)$, where b is the semi-minor axis of the ellipse, γ is the sign of circular polarization, and γ = 1 and γ = −1 denote the rotation of the electric field vector to the right and to the left, respectively.

Expression (14) makes it possible to calculate the angle ψ, as well as the length of the minor b and major a semiaxes of the refracted beam polarization ellipse using the following Equations:

$$\psi = \frac{1}{2}\tan^{-1}\left(\frac{2T_{x''}T_y\cos\delta_t}{T_{x''}{}^2 - T_y{}^2}\right), \tag{15}$$

$$a = \sqrt{\frac{1}{2}\left(T_{x''}{}^2 + T_y{}^2 + \sqrt{T_{x''}{}^4 + T_y{}^4 + 2T_{x''}{}^2 T_y{}^2\cos 2\delta_t}\right)}, \tag{16}$$

$$b = \sqrt{\frac{1}{2}\left(T_{x''}{}^2 + T_y{}^2 - \sqrt{T_{x''}{}^4 + T_y{}^4 + 2T_{x''}{}^2 T_y{}^2\cos 2\delta_t}\right)}. \tag{17}$$

Equations (16) and (17) were used to calculate P_{cir} and P_{lin}. Figure 6a,b shows the calculated values of ψ, P_{cir}, and P_{lin} for different angles α, for which the waveforms of the photovoltage pulses were recorded for circular and linear polarizations (Φ = −45°) of the incident radiation, respectively. It can be seen that when the incident beam is circularly polarized at large angles α, there is a significant change in the parameters ψ, P_{cir}, and P_{lin} characterizing the polarization ellipse in the refracted beam. However, at α < 20°, the refracted beam remains virtually circularly polarized, i.e., at small angles of incidence, the polarizations of the incident and refracted beams virtually coincide. It should be added that at all α, the signs of the circular polarization of the incident and refracted beams coincide, i.e., for both beams $\gamma = 1$.

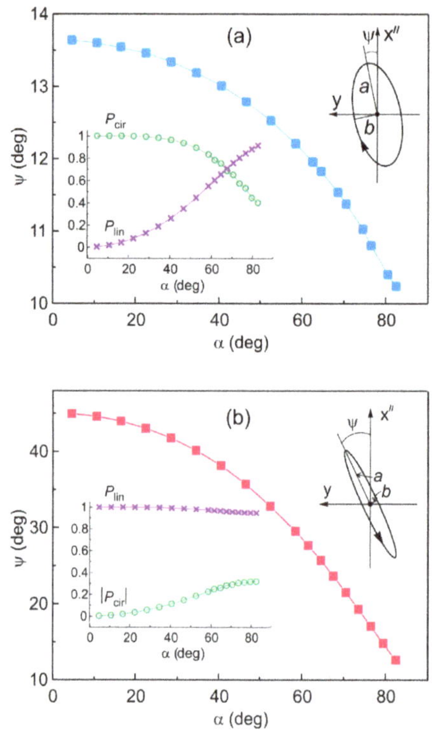

Figure 6. The angle ψ of the semi-major axis a of the refracted beam polarization ellipse relative to the plane of incidence σ, as well as the degrees of circular P_{cir} and linear P_{lin} polarizations of the refracted beam (insets) as a function of the angle of incidence α for (**a**) circularly and (**b**) linearly polarized ($\Phi = -45°$) radiation incident on a thin CuSe/t-Se film (points are calculated values of ψ, P_{cir} and P_{lin} for angles α, for which photovoltage pulses were recorded, curves denote smoothing functions). The graphical insets show the corresponding calculated polarization ellipses of the refracted beam at $\alpha = 64.5°$ in the x″y coordinate system, where the x″ axis lies in the refraction plane coinciding with the plane of incidence σ and is perpendicular to the wave vector of the refracted beam.

It follows from Figure 6b that with linear polarization of the incident light at $\Phi = -45°$ the refracted beam is also elliptically polarized, but in this case $\gamma = -1$ (see upper inset). In addition, when changing α in the range of 0–85°, the modulus P_{cir} of the refracted beam changes insignificantly from 0 to 0.32. At small angles α, the changes in the polarization state of the refracted beam are minimal.

It is known that the transverse photocurrent j_y for elliptical excitation beam polarization consists of a CPC ($j_{y,cir}$) and an LPC ($j_{y,lin}$), i.e., $j_y = j_{y,cir} + j_{y,lin}$. According to [48] and taking into account [29], for positive angles of incidence, the transverse photocurrent amplitude in the CuSe/Se film structure for elliptical beam polarization can be represented as follows:

$$j_y = A_{cir}P_{cir} - A_{lin}P_{lin}\sin 2\psi, \quad (18)$$

where A_{cir} and A_{lin} are the positive coefficients of the circular and linear photocurrents at a given α, respectively. Taking into account Equation (18) and in accordance with the results of [37], the waveforms of the transverse photovoltage pulses $U_y(t, \alpha)$ can be written as follows:

$$U_y(t, \alpha) = B_{cir}(\alpha)P_{cir}f_{cir}(t) - B_{lin}(\alpha)P_{lin}\sin 2\psi\, f_{lin}(t), \quad (19)$$

where $B_{cir}(\alpha)$, $B_{lin}(\alpha)$ are positive coefficients (for positive α) depending on α and characterizing the circular and linear contributions, respectively, and $f_{cir}(t)$ and $f_{lin}(t)$ are the transverse photovoltage waveforms normalized to the maximum values, recorded at angles α close to zero for circular and linear polarizations, respectively. Equation (19) allows one to approximate the waveforms of the photovoltage pulses, recorded at different α, with two unknown parameters $B_{cir}(\alpha)$ and $B_{lin}(\alpha)$. For example, Figure 7 shows the approximations of the waveforms for four of the recorded pulses at different α. It can be seen that the obtained approximating curves satisfactorily describe the experiment. It should be noted that if the waveforms $f_{cir}(t)$ and $f_{lin}(t)$ coincide with each other, the generation of bipolar pulses is impossible, and the duration of unipolar pulses of the resulting photocurrent, defined by Equation (19), does not depend on the angle of incidence.

Figure 8 shows the dependences of the calculated coefficients B_{cir} and B_{lin} on α. It follows from the figure that both coefficients B_{cir} and B_{lin} approach zero at small and also at grazing angles of incidence. However, the dependences of B_{cir} and B_{lin} on α differ significantly from each other. It can be seen that the dependences of B_{cir} and B_{lin} acquire their extreme values at $\alpha = 54°$ and $\alpha = 62°$, respectively. The coefficient B_{lin} prevails over B_{cir} for the entire range of the incident angle change and the ratio B_{lin}/B_{cir} increases monotonically with increasing α.

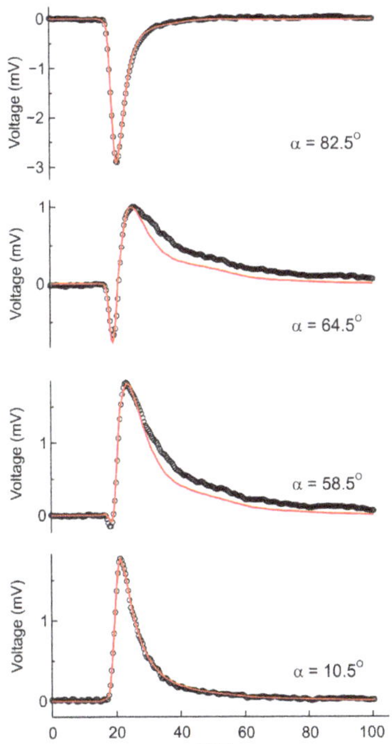

Figure 7. The oscillograms of the photovoltage pulses arising in a thin CuSe/*t*-Se film with circular polarization of exciting radiation and angles of incidence α = 10.5, 58.5, 64.5, and 82.5° (black circles), and their approximations shown by red lines according to Equation (19).

It follows from Figure 6b that with linear polarization ($\Phi = -45°$) of the incident beam, the refracted beam becomes elliptically polarized at a negative polarization sign ($\gamma = -1$). This means that both terms on the right side of Equation (19) remain negative for any positive angle of incidence. In addition, when changing α in the range of 0–85°, the parameters P_{cir} and P_{lin} of the refracted beam do not change significantly. Taking into account Equation (19), this leads to a weak dependence of the waveform of the photovoltage pulse on the angle α for linearly polarized laser pumping. It should be noted that the transverse photocurrent arising in the medium due to the PDE or CPGE also consists of the circular and linear contributions. Therefore, the temporal profile of the transverse photocurrent pulses of the PDE or CPGE generated in the nonlinear optical medium when pumped by short laser pulses of circular polarization can also depend in a complex way on the angle at which the light falls on the surface of the absorbing medium.

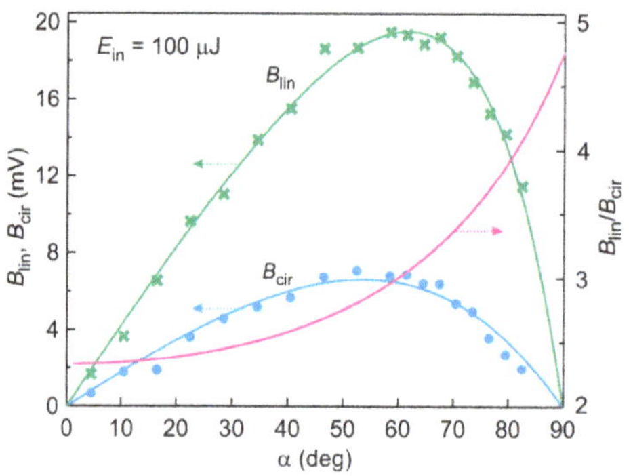

Figure 8. The coefficients B_{lin} (green crosses), B_{cir} (blue dots) as a function of the angle of incidence α, which characterize the linear and circular contributions to the transverse photovoltage pulses, calculated from the experimental data (the green and blue curves represent the corresponding approximations by the equations $B_{lin} = 55.98 \sin 2\alpha/(1.15\cos\alpha + 1)^2$, $B_{cir} = 11.75 \sin 2\alpha\ /(0.5\cos\alpha + 1)^2$, respectively), and calculated dependence of the ratio B_{lin}/B_{cir} on α (pink curve).

Good agreement between the experimental data and the calculated dependences confirms that the angular dependence of the waveforms of transverse photocurrent pulses in a thin CuSe/Se film under circularly polarized femtosecond laser pumping originates from the transformation of circular polarization into elliptical polarization upon refraction of light in a semitransparent CuSe/Se film and the interaction of LPC and CPC having different relaxation times in the film structure. As mentioned above, despite the large number of publications on the topic of generation of polarization-sensitive transverse photocurrent in various materials, arising by various mechanisms, such a phenomenon has not been observed before. It is possible that this is due to the fact that in many studies of the transverse photocurrent, cw laser radiation or nanosecond laser radiation has been used (see, for example, [33,67–69]).

The results obtained in this work can be used in various applications in optoelectronics. For example, the experimental setup presented in the inset to Figure 5 can be used for the fast direct detection of the circular polarization state of light. If at angles of incidence $0 < \alpha < 58.5°$ (for example, at $\alpha = 45°$) the photocurrent pulses generated in a CuSe/*t*-Se thin film have a positive polarity, then this means that the incident radiation is right-hand polarized (looking towards the light source). If, under the same experimental conditions, the photocurrent pulses have a negative polarity, then the incident radiation is left-hand polarized. This method of determining the state of circular polarization does not require the use of optical elements. Further, it is obvious that this photovoltaic property of a CuSe/Se thin film can be used to determine the fast and slow axes of a quarter-wave plate without using a reference quarter-wave plate and an optical light polarization.

4. Conclusions

In thin semitransparent CuSe/*t*-Se films synthesized by vacuum thermal deposition, the generation of nanosecond photocurrent pulses is studied as a function of the angle of incidence and polarization of exciting femtosecond laser pulses at 795 nm. It has been established that the dependences of the longitudinal and transverse photocurrents on the angle of incidence are described by odd relationships, which are characteristic of the SPGE nonlinear optical phenomenon. The relationships found for the longitudinal and transverse photocurrents as a function of pump polarization are also in agreement with the mechanism of photocurrent generation due to the SPGE. It is shown that the pulse durations of the longitudinal photocurrent for linear and circular polarizations, as well as the transverse photocurrent for linear polarization, are virtually independent of the incidence angle. However, the waveforms of the transverse photocurrent pulses at circular polarization with a given direction of rotation of the electric field vector of the incident radiation at a fixed sign of the angle of incidence strongly depend on the angle of incidence. At small and large angles of incidence, unipolar pulses of opposite polarity are generated, and in the intermediate range of incidence angles ($58.5 \leq \alpha \leq 76.5^\circ$), bipolar photocurrent pulses are excited, smoothly transforming into unipolar pulses of opposite polarity at the boundaries of this interval. The obtained features of the waveform of the transverse photocurrent pulses at circular polarization of the incident radiation are due to the following: (i) the transformation of the circular polarization of the incident radiation into an elliptical one (without changing the sign of the circular polarization) upon the refraction of light at the air/semitransparent film interface and the appearance of a linear component of the photocurrent in the film structure, depending on the angle of incidence; (ii) the interaction of multidirectional linear and circular components of the photocurrent, which have different relaxation times and strongly depend on the angle of incidence.

The relationships found for the influence of the incidence angle on the waveform of pulses of longitudinal and transverse photocurrents that arise in the CuSe/*t*-Se film structure under polarized pulsed pumping due to the SPGE can be found in various materials in which the PSPC is excited by the PDE or by the CPGE.

The results obtained in this work can be used in optoelectronics, in particular, to create a high-speed detector capable of distinguishing left-handed and right-handed polarized light, as well as in the development of a technique that allows one to quickly determine the fast and slow axes of quarter-wave plates.

Supplementary Materials: The following supporting information can be downloaded at: https://www.mdpi.com/article/10.3390/app12146869/s1, Figure S1: X-ray diffractogram of the CuSe/*t*-Se nanocomposite film and diffraction patterns of CuSe (PDF 00-034-0171) and *t*-Se (PDF 00-042-1425) powders. A copper-based X-ray tube generating radiation at a wavelength of 0.1541 nm was used; Figure S2: Raman spectrum of the CuSe/*t*-Se nanocomposite film (black line). The red line represents the fitting of the combined Gaussian profiles at (blue line) 232 and (purple line) 237 cm^{-1} corresponding to the *t*-Se and (green line) 262 cm^{-1} to the CuSe nanocrystallites Raman resonances. A He-Ne laser radiation at a wavelength of 632.8 nm as excitation pumping was used; Figure S3: Scanning electron microscope image of the CuSe/*t*-Se semitransparent thin film surface; Figure S4: The optical transmittance spectrum of the CuSe/*t*-Se nanocomposite.

Author Contributions: Conceptualization, A.E.F. and G.M.M.; methodology, A.E.F., V.Y.K. and G.M.M.; validation, A.E.F., T.N.M., K.G.M. and G.M.M.; investigation, A.E.F., T.N.M., V.Y.K., K.G.M. and G.M.M.; resources, V.Y.K. and G.M.M.; writing—original draft preparation, A.E.F. and G.M.M.; writing—review and editing, all authors; supervision, G.M.M.; project administration, G.M.M. All authors have read and agreed to the published version of the manuscript.

Funding: This work was supported by the Ministry of Education and Science of the Russian Federation (state registration number 1021032422167-7-1.3.2) and the Academy of Finland (Grant Nos. 323053, 340115).

Institutional Review Board Statement: Not applicable.

Informed Consent Statement: Not applicable.

Data Availability Statement: Not applicable.

Acknowledgments: The authors are grateful to Y.P. Svirko for discussing the results obtained. This study was performed using equipment of the Shared Use Center "Center of Physical and Physicochemical Methods of Analysis and Study of the Properties and Surface Characteristics of Nanostructures, Materials, and Products" UdmFRC UB RAS.

Conflicts of Interest: The authors declare no conflict of interest.

Nomenclature

Symbol	
α	Angle of incidence
Se	Selenium
Cu	Copper
t-Se	Trigonal selenium
E_{in}	Laser pulse energy
R_{in}	Input impedance of the oscilloscope
j_x	Longitudinal photocurrent
j_y	Transverse photocurrent
U_x	Extreme values of the voltage pulses at the experimental geometry when the measuring electrodes were placed perpendicular to the plane of incidence
U_y	Extreme values of the voltage pulses at the experimental geometry when the measuring electrodes were placed parallel to the plane of incidence
t	Time
η_x	Efficiency of converting light into longitudinal photocurrent
η_y	Efficiency of converting light into transverse photocurrent
Φ	Angle between the plane of polarization and the radiation incidence plane on the film
σ	Radiation incidence plane on the film
$\mathbf{n}$	Normal to the film surface
$\mathbf{A}$	Measuring electrode
$\mathbf{B}$	Measuring electrode
$\mathbf{k}$	The wave vector of the optical field
$\mathbf{E}$	Electric field vector of the incident radiation
$\mathbf{E}^{(\mathbf{t})}$	Electric field vector of the refracted beam
x, y	Axes of the rectangular coordinate system
x′	Axis, which lies in the σ plane and perpendicular to $\mathbf{k}$
x″	Axis, which lies in the refraction plane coinciding with the plane of incidence and perpendicular to the wave vector of the refracted beam
t_p	Complex Fresnel refractive index for *p*-polarizations
t_s	Complex Fresnel refractive index for *s*-polarizations
$\hat{n}$	Complex refractive index
n	Real refractive index that determines the phase velocity
κ	Absorption coefficient
δ_t	Phase shift between the components of the refracted beam with *p*- and *s*-polarizations
δ_{tp}	Phase shift of the *p*-component of the beam, resulting from refraction at the interface between two media with a complex refractive index
δ_{ts}	Phase shift of the *s*-component of the beam, resulting from refraction at the interface between two media with a complex refractive index
δ_0	Initial phase shift between the *p*- and *s*-components before refraction
a	Semi-major axis of the refracted beam polarization ellipse
b	Semi-minor axis of the refracted beam polarization ellipse
P_{cir}	Degree of circular polarization
P_{lin}	Degree of linear polarization
ψ	Angle between the semi-major axis of the refracted beam polarization ellipse and axis x″
γ	Sign of circular polarization
$U_y(\mathrm{t}, \alpha)$	Waveforms of the transverse photovoltage pulses
A_{cir}	Positive coefficient of the circular photocurrent at a given α
A_{lin}	Positive coefficient of the linear photocurrent at a given α
B_{cir}	Positive coefficient (for positive α) depending on α and characterizing the circular contribution
B_{lin}	Positive coefficient (for positive α) depending on α and characterizing the linear contribution
$f_{cir}(t)$	Transverse photovoltage waveform normalized to its maximum value, recorded at angle α close to zero for circular polarizations
$f_{lin}(t)$	Transverse photovoltage waveform normalized to its maximum values, recorded at angle α close to zero for linear polarizations

Abbreviations

PSPC	Polarization sensitive photocurrent
CPC	Circular photocurrent
LPC	Linear photocurrent
CPGE	Circular photogalvanic effect
PGE	Photogalvanic effect
PDE	Photon drag effect
SEM	Scanning electron microscope
FWHM	Full width at half maximum

References

1. Wang, J.; Zhou, Y.J.; Xiang, D.; Ng, S.J.; Watanabe, K.; Taniguchi, T.; Eda, G. Polarized light-emitting diodes based on anisotropic excitons in few-layer ReS2. *Adv. Mater.* **2020**, *32*, 2001890. [CrossRef] [PubMed]
2. Wang, X.; Wang, Q.; Zhang, X.; Miao, J.; Cheng, J.; He, T.; Li, Y.; Tang, Z.; Chen, R. Circularly polarized light source from self-assembled hybrid nanoarchitecture. *Adv. Opt. Mater.* **2022**, *2022*, 2200761. [CrossRef]
3. Seo, I.C.; Lim, Y.; An, S.C.; Woo, B.H.; Kim, S.; Son, J.G.; Yoo, S.; Park, Q.H.; Kim, J.Y.; Jun, Y.C. Circularly polarized emission from organic-inorganic hybrid perovskites via chiral fano resonances. *ACS Nano* **2021**, *15*, 13781–13793. [CrossRef] [PubMed]
4. Ivchenko, E.L. *Optical Spectroscopy of Semiconductor Nanostructures*; Springer: Berlin/Heidelberg, Germany, 2004.
5. Gurevich, V.L.; Laiho, R. Photomagnetism of metals. First observation of dependence on polarization of light. *Phys. Solid State* **2000**, *42*, 1807–1812. [CrossRef]
6. Singh, A.; Li, X.; Protasenko, V.; Galantai, G.; Kuno, M.; Xing, H.; Jena, D. Polarization-sensitive nanowire photodetectors based on solution-synthesized CdSe quantum-wire solids. *Nano Lett.* **2007**, *7*, 2999–3006. [CrossRef] [PubMed]
7. Luo, Y.; Hu, Y.; Xie, Y. Highly polarization-sensitive, visible-blind and self-powered ultraviolet photodetection based on two-dimensional wide bandgap semiconductors: A theoretical prediction. *J. Mater. Chem. A* **2019**, *7*, 27503–27513. [CrossRef]
8. Zhao, Q.; Gao, F.; Chen, H.; Gao, W.; Xia, M.; Pan, Y.; Shi, H.; Su, S.; Fang, X.; Li, J. High performance polarization-sensitive self-powered imaging photodetectors based on a p-Te/n-MoSe2 van der Waals heterojunction with strong interlayer transition. *Mater. Horiz.* **2021**, *8*, 3113–3123. [CrossRef]
9. Qian, L.; Zhao, J.; Xie, Y. Enhanced photogalvanic effect in the two-dimensional MgCl2/ZnBr2 vertical heterojunction by inhomogenous tensile stress. *Front. Phys.* **2022**, *17*, 13502. [CrossRef]
10. Karch, J.; Olbrich, P.; Schmalzbauer, M.; Zoth, C.; Brinsteiner, C.; Fehrenbacher, M.; Wurstbauer, U.; Glazov, M.M.; Tarasenko, S.A.; Ivchenko, E.L.; et al. Dynamic Hall effect driven by circularly polarized light in a graphene layer. *Phys. Rev. Lett.* **2010**, *105*, 227402. [CrossRef]
11. Ganichev, S.D.; Ivchenko, E.L.; Prettl, W. Photogalvanic effects in quantum wells. *Phys. E Low-Dimens. Syst. Nanostruct.* **2002**, *14*, 166–171.
12. Pan, Y.; Wang, Q.Z.; Yeats, A.L.; Pillsbury, T.; Flanagan, T.C.; Richardella, A.; Zhang, H.; Awschalom, D.D.; Liu, C.X.; Samarth, N. Helicity dependent photocurrent in electrically gated (Bi1-x Sb x)2Te3 thin films. *Nat. Commun.* **2017**, *8*, 1037. [CrossRef]
13. Zhang, Z.; Zhang, R.; Liu, B.; Xie, Z.L.; Xiu, X.Q.; Han, P.; Lu, H.; Zheng, Y.D.; Chen, Y.H.; Tang, C.G.; et al. Circular photogalvanic effect at inter-band excitation in InN. *Solid State Commun.* **2008**, *145*, 159–162. [CrossRef]
14. Mikheev, G.M.; Kogai, V.Y.; Mikheev, K.G.; Mogileva, T.N.; Saushin, A.S.; Svirko, Y.P. Polarization-sensitive photoresponse of the CuSe/Se nanocomposite prepared by vacuum thermal deposition. *Mater. Today Commun.* **2019**, *21*, 100656. [CrossRef]
15. Hägele, D.; Oestreich, M.; Rühle, W.W.; Nestle, N.; Eberl, K. Spin transport in GaAs. *Appl. Phys. Lett.* **1998**, *73*, 1580–1582. [CrossRef]
16. Okada, K.N.; Ogawa, N.; Yoshimi, R.; Tsukazaki, A.; Takahashi, K.S.; Kawasaki, M.; Tokura, Y. Enhanced photogalvanic current in topological insulators via Fermi energy tuning. *Phys. Rev. B* **2016**, *93*, 081403(R). [CrossRef]
17. Yu, J.; Zhu, K.; Zeng, X.; Chen, L.; Chen, Y.; Liu, Y.; Yin, C.; Cheng, S.; Lai, Y.; Huang, J.; et al. Helicity-dependent photocurrent of the top and bottom Dirac surface states of epitaxial thin films of three-dimensional topological insulators Sb2Te3. *Phys. Rev. B* **2019**, *100*, 235108. [CrossRef]
18. Mikheev, G.M.; Styapshin, V.M. Nanographite analyzer of laser polarization. *Instrum. Exp. Tech.* **2012**, *55*, 85–89. [CrossRef]
19. Akbari, M.; Ishihara, T. Polarization dependence of transverse photo-induced voltage in gold thin film with random nanoholes. *Opt. Express* **2017**, *25*, 2143–2152. [CrossRef]
20. Roy, S.; Manna, S.; Mitra, C.; Pal, B. Photothermal Control of Helicity-Dependent Current in Epitaxial Sb_2Te_2Se Topological Insulator Thin-Films at Ambient Temperature. *ACS Appl. Mater. Interfaces* **2022**, *14*, 9909–9916. [CrossRef]
21. Mirzaee, S.M.A.; Lebel, O.; Nunzi, J.-M. A simple unbiased hot-electron polarization-sensitive near- infrared photo-detector. *ACS Appl. Mater. Interfaces* **2018**, *10*, 11862–11871. [CrossRef]
22. Li, W.; Coppens, Z.J.; Besteiro, L.V.; Wang, W.; Govorov, A.O.; Valentine, J. Circularly polarized light detection with hot electrons in chiral plasmonic metamaterials. *Nat. Commun.* **2015**, *6*, 8379. [CrossRef]
23. Schulz, M.; Balzer, F.; Scheunemann, D.; Arteaga, O.; Lützen, A.; Meskers, S.C.J.; Schiek, M. Chiral excitonic organic photodiodes for direct detection of circular polarized light. *Adv. Funct. Mater.* **2019**, *29*, 1900684. [CrossRef]

24. Wang, L.; Xue, Y.; Cui, M.; Huang, Y.; Xu, H.; Qin, C.; Yang, J.; Dai, H.; Yuan, M. A chiral reduced-dimension perovskite for an efficient flexible circularly polarized light photodetector. *Angew. Chemie* **2020**, *132*, 6504–6512. [CrossRef]
25. Hao, J.; Lu, H.; Mao, L.; Chen, X.; Beard, M.C.; Blackburn, J.L. Direct detection of circularly polarized light using chiral copper chloride-carbon nanotube heterostructures. *ACS Nano* **2021**, *15*, 7608–7617. [CrossRef] [PubMed]
26. Belinicher, V.I.; Sturman, B.I. The photogalvanick effect in media lacking of a center of symmetry. *Sov. Phys. Uspekhi* **1980**, *23*, 199–223. [CrossRef]
27. Ganichev, S.D.; Prettl, W.J. Spin photocurrents in quantum wells. *Phys. Condens. Matter.* **2003**, *15*, R935–R983. [CrossRef]
28. Zhang, Y.; Cao, R.; Hu, Y.; Wang, Y.; Xie, Y. A promising polarization-sensitive ultraviolet photodetector based on the two-dimensional ZrNBr-ZrNCl lateral heterojunction with enhanced photoresponse: A theoretical prediction. *Appl. Surf. Sci.* **2021**, *560*, 149907. [CrossRef]
29. Ivchenko, E.L. Circular photogalvanic effect in nanostructures. *Physics-Uspekhi* **2002**, *45*, 1299–1303. [CrossRef]
30. Ivchenko, E.L.; Spivak, B. Circular photogalvanic effect and related effects in chiral carbon nanotubes. *Physica* **2003**, *17*, 376–379. [CrossRef]
31. Dhara, S.; Mele, E.J.; Agarwal, R. Voltage-tunable circular photogalvanic effect in silicon nanowires. *Science* **2015**, *349*, 726–729. [CrossRef]
32. Moayed, M.M.R.; Li, F.; Beck, P.; Schober, J.-C.; Klinke, C. Anisotropic circular photogalvanic effect in colloidal tin sulfide nanosheets. *Nanoscale* **2020**, *12*, 6256–6262. [CrossRef]
33. Hubmann, S.; Budkin, G.V.; Otteneder, M.; But, D.; Sacré, D.; Yahniuk, I.; Diendorfer, K.; Bel'kov, V.V.; Kozlov, D.A.; Mikhailov, N.N.; et al. Symmetry breaking and circular photogalvanic effect in epitaxial CdxHg1-xTe films. *Phys. Rev. Mater.* **2020**, *4*, 043607. [CrossRef]
34. Sun, X.; Adamo, G.; Eginligil, M.; Krishnamoorthy, H.N.S.; Zheludev, N.I.; Soci, C. Topological insulator metamaterial with giant circular photogalvanic effect. *Sci. Adv.* **2021**, *7*, eabe5748. [CrossRef] [PubMed]
35. Glazov, M.M.; Ganichev, S.D. High frequency electric field induced nonlinear effects in graphene. *Phys. Rep.* **2014**, *535*, 101–138. [CrossRef]
36. Mikheev, G.M.; Saushin, A.S.; Vanyukov, V.V.; Mikheev, K.G.; Svirko, Y.P. Femtosecond circular photon drag effect in the Ag/Pd nanocomposite. *Nanoscale Res. Lett.* **2017**, *12*, 39. [CrossRef] [PubMed]
37. Akbari, M.; Onoda, M.; Ishihara, T. Photo-induced voltage in nano-porous gold thin film. *Opt. Express* **2015**, *23*, 823–832. [CrossRef]
38. Khichar, V.; Sharma, S.C.; Hozhabri, N. New features in the surface plasmon induced photon drag effect in noble metal thin films. *J. Phys. Commun.* **2021**, *5*, 055005. [CrossRef]
39. Al'perovich, V.L.; Belinicher, V.I.; Novikov, V.N.; Terekhov, A.S. Surface photovoltaic effect in solids. Theory and experiment for interband transitions in gallium arsenide. *Sov. Phys. JETP* **1981**, *53*, 1201–1208.
40. Mikheev, G.M.; Saushin, A.S.; Styapshin, V.M.; Svirko, Y.P. Interplay of the photon drag and the surface photogalvanic effects in the metal-semiconductor nanocomposite. *Sci. Rep.* **2018**, *8*, 8644. [CrossRef]
41. Guo, B.; Xiao, Q.L.; Wang, S.H.; Zhang, H. 2D Layered materials: Synthesis, nonlinear optical properties, and device applications. *Laser Photonics Rev.* **2019**, *13*, 1800327. [CrossRef]
42. You, J.W.; Bongu, S.R.; Bao, Q.; Panoiu, N.C. Nonlinear optical properties and applications of 2D materials: Theoretical and experimental aspects. *Nanophotonics* **2018**, *8*, 63–97. [CrossRef]
43. Gibson, A.F.; Kimmitt, M.F.; Walker, A.C. Photon drag in germanium. *Appl. Phys. Lett.* **1970**, *17*, 75–77. [CrossRef]
44. Noginova, N.; Rono, V.; Bezares, F.J.; Caldwell, J.D. Plasmon drag effect in metal nanostructures. *New J. Phys.* **2013**, *15*, 113061. [CrossRef]
45. Mikheev, G.M.; Saushin, A.S.; Vanyukov, V.V. Helicity-dependent photocurrent in the resistive Ag/Pd films excited by IR laser radiation. *Quantum Electron.* **2015**, *45*, 635–639. [CrossRef]
46. Strait, J.H.; Holland, G.; Zhu, W.; Zhang, C.; Ilic, B.R.; Agrawal, A.; Pacifici, D.; Lezec, H.J. Revisiting the photon-drag effect in metal films. *Phys. Rev. Lett.* **2019**, *123*, 53903. [CrossRef] [PubMed]
47. Beregulin, E.V.; Valov, P.M.; Ryvkin, S.M.; Yaroshetskii, I.D.; Lisker, I.S.; Pukshanskii, A.L. Dragging of electrons by light in semimetals. *JETP Lett.* **1977**, *25*, 101–104.
48. Mikheev, G.M.; Fateev, A.E.; Kogai, V.Y.; Mogileva, T.N.; Vanyukov, V.V.; Svirko, Y.P. Helicity dependent temporal profile of the semiconductor thin film photoresponse. *Appl. Phys. Lett.* **2021**, *118*, 201105. [CrossRef]
49. Hirose, H.; Ito, N.; Kawaguchi, M.; Lau, Y.C.; Hayashi, M. Circular photogalvanic effect in Cu/Bi bilayers. *Appl. Phys. Lett.* **2018**, *113*, 222404. [CrossRef]
50. Saushin, A.S.; Mikheev, K.G.; Styapshin, V.M.; Mikheev, G.M. Direct measurement of the circular photocurrent in the Ag/Pd nanocomposites. *J. Nanophotonics* **2017**, *11*, 032508. [CrossRef]
51. Konchenkov, V.I.; Myachkova, A.A.; Zav'Yalov, D.V. Influence of a constant field on a circular photovoltaic effect in two-dimensional superlattices. *J. Phys. Conf. Ser.* **2020**, *1697*. [CrossRef]
52. Mirzaee, S.M.A.; Nunzi, J.-M. Searching for evidence of optical rectification: Optically induced nonlinear photovoltage in a capacitor configuration. *J. Opt. Soc. Am. B* **2019**, *36*, 53. [CrossRef]
53. Aleksandrovich, E.V.; Aleksandrovich, A.N.; Mikheev, G.M. Laser-induced modification of optical properties of glassy selenium films synthesized by vacuum thermal evaporation. *J. Non. Cryst. Solids* **2020**, *545*, 120249. [CrossRef]

54. Li, Q.; Qi, D.; Wang, X.; Shen, X.; Wang, R.; Tanaka, K. Femto- and nano-second laser-induced damages in chalcogenide glasses. *Jpn. J. Appl. Phys.* **2019**, *58*, 080911. [CrossRef]
55. Ambade, S.B.; Mane, R.S.; Kale, S.S.; Sonawane, S.H.; Shaikh, A.V.; Han, S.-H. Chemical synthesis of p-type nanocrystalline copper selenide thin films for heterojunction solar cells. *Appl. Surf. Sci.* **2006**, *253*, 2123–2126. [CrossRef]
56. Gosavi, S.R.; Deshpande, N.G.; Gudage, Y.G.; Sharma, R. Physical, optical and electrical properties of copper selenide (CuSe) thin films deposited by solution growth technique at room temperature. *J. Alloys Compd.* **2008**, *448*, 344–348. [CrossRef]
57. Hankare, P.P.; Khomane, A.S.; Chate, P.A.; Rathod, K.C.; Garadkar, K.M. Preparation of copper selenide thin films by simple chemical route at low temperature and their characterization. *J. Alloys Compd.* **2009**, *469*, 478–482. [CrossRef]
58. Yakuphanoglu, F.; Viswanathan, C. Electrical conductivity and single oscillator model properties of amorphous CuSe semiconductor thin film. *J. Non. Cryst. Solids* **2007**, *353*, 2934–2937. [CrossRef]
59. Gao, L.; Sun, J.T.; Lu, J.C.; Li, H.; Qian, K.; Zhang, S.; Zhang, Y.Y.; Qian, T.; Ding, H.; Lin, X.; et al. Epitaxial growth of honeycomb monolayer CuSe with Dirac nodal line fermions. *Adv. Mater.* **2018**, *30*, 1707055. [CrossRef]
60. Jadhav, C.D.; Rondiya, S.R.; Hambire, R.C.; Baviskar, D.R.; Deore, A.V.; Cross, R.W.; Dzade, N.Y.; Chavan, P.G. Highly efficient field emission properties of vertically aligned 2D CuSe nanosheets: An experimental and theoretical investigation. *J. Alloys Compd.* **2021**, *875*, 159987. [CrossRef]
61. Ho, S.M. Fabrication of Cu 4 SnS 4 Thin Films: A Review. *Eng. Technol. Appl. Sci. Res.* **2020**, *10*, 6161–6164. [CrossRef]
62. Shinde, M.S.; Ahirrao, P.B.; Patil, I.J.; Patil, R.S. Thickness dependent electrical and optical properties of nanocrystalline copper sulphide thin films grown by simple chemical route. *Indian J. Pure Appl. Phys.* **2012**, *50*, 657–660.
63. Ezenwa, I.A.; Okereke, N.A.; Okoli, L.N. Electrical properties of copper selenide thin film. *IPASJ Int. J. Electr. Eng.* **2013**, *1*, 1–4.
64. García, V.M.; Nair, M.T.S.; Nair, P.K.; Zingaro, R.A. Chemical deposition of bismuth selenide thin films using N,N-dimethylselenourea. *Semicond. Sci. Technol.* **1997**, *12*, 645–653. [CrossRef]
65. Mikheev, G.M.; Kogai, V.Y.; Mikheev, K.G.; Mogileva, T.N.; Saushin, A.S.; Svirko, Y.P. Interaction of the polarization-sensitive surface photocurrents in the semitransparent CuSe/Se film. *Opt. Express* **2021**, *29*, 2112–2123. [CrossRef]
66. Saushin, A.S.; Mikheev, G.M.; Vanyukov, V.V.; Svirko, Y.P. The surface photogalvanic and photon drag effects in Ag/Pd metal-semiconductor nanocomposite. *Nanomaterials* **2021**, *11*, 2827. [CrossRef] [PubMed]
67. Zonov, R.G.; Mikheev, G.M.; Obraztsov, A.N.; Svirko, Y.P. Circular photocurrent in the carbon nanowall film. *Opt. Lett.* **2020**, *45*, 2022–2025. [CrossRef]
68. Wang, S.; Zhang, H.; Zhang, J.; Li, S.; Luo, D.; Wang, [illegible].; Jin, K.; Sun, J. Circular Photogalvanic Effect in Oxide Two-Dimensional Electron Gases. *Phys. Rev. Lett.* **2022**, *128*, 187401. [CrossRef]
69. Li, M.; Yu, J.; Cui, G.; Chen, Y.; Lai, Y.; Cheng, S.; He, K. Circular photogalvanic effect of surface states in the topological insulator Bi2(Te0.23Se0.77)3 nanowires grown by chemical vapor deposition. *J. Appl. Phys.* **2022**, *131*, 113902. [CrossRef]

applied sciences

Article

Edge Currents Induced by AC Electric Field in Two-Dimensional Dirac Structures

Mikhail V. Durnev and Sergey A. Tarasenko *

Ioffe Institute, Politechnicheskaya 26, 194021 St. Petersburg, Russia
* Correspondence: tarasenko@coherent.ioffe.ru

Abstract: Edges in two-dimensional structures are the source of nonlinear transport and optical phenomena which are particularly important in small-size flakes. We present a microscopic theory of the edge photogalvanic effect, i.e., the formation of DC electric current flowing along the sample edges in response to AC electric field of the incident terahertz radiation, for two-dimensional Dirac materials including the systems with massive and massless charge carriers. The edge current direction is controlled by the AC field polarization. The spectral dependence of the current is determined by the carrier dispersion and the mechanism of carrier scattering, as shown for single-layer and bilayer graphene as examples.

Keywords: edge currents; high-frequency nonlinear transport; photogalvanic effect; two-dimensional Dirac structures; massive and massless fermions

Citation: Durnev, M.V.; Tarasenko, S.A. Edge Currents Induced by AC Electric Field in Two-Dimensional Dirac Structures. *Appl. Sci.* **2023**, *13*, 4080. https://doi.org/10.3390/app13074080

Academic Editor: Gennady M. Mikheev

Received: 25 February 2023
Revised: 19 March 2023
Accepted: 20 March 2023
Published: 23 March 2023

1. Introduction

The discovery of graphene and other two-dimensional (2D) crystals opened a new page in the physics of low-dimensional systems [1,2] and triggered the research aimed at the development of efficient sources and detectors of terahertz radiation based on 2D Dirac materials [3,4]. In small-size samples, e.g., flakes obtained by mechanical exfoliation, the important and sometimes decisive role in the formation of photoelectric response is played by edges and nearby regions [5,6]. At the edges, the translational and space inversion symmetries are naturally broken, which gives rise to edge-related mechanisms of the photogalvanic effect [5–11] and the second harmonic generation [12–14].

The photocurrents flowing along the sample edges (the edge photogalvanic effect) were observed and studied in single-layer and bilayer graphene samples excited by terahertz radiation [5,6,10], also in an external magnetic field in the conditions of cyclotron resonance [9] and in the regime of the quantum Hall effect [15]. It is found that the edge photocurrent is induced by both linearly and circularly polarized radiation. Moreover, the photocurrent direction is controlled by the polarization of the incident radiation: the electric field direction with respect to the edge for the linearly polarized field and the photon helicity for the circularly polarized field. The edge photogalvanic effect in 2D structures can be considered as a low-dimensional analog of the surface photogalvanic effect studied in bulk semiconductor crystals and metal films and recently in nanocomposite films [16–22].

The microscopic theory of the edge photogalvanic effect in the spectral range of intraband transport has been developed so far for 2D systems with parabolic energy spectrum of charge carriers [6,8]. Here, we generalize the theory to the class of 2D Dirac materials. We present a comprehensive theoretical study of the edge currents for 2D systems with arbitrary dispersion $\varepsilon(p)$ and arbitrary type of electron scattering in the 2D bulk. We show that the edge current is determined by the dispersion of carriers and the relaxation times of the first and second angular harmonics of the distribution function and compare the results for single-layer and bilayer graphene, which are examples of 2D systems with parabolic and linear energy spectra.

2. Microscopic Theory

Consider an electromagnetic wave incident on the structure hosting a 2D electron gas occupying the (xy) half-plane at $x \geq 0$, see Figure 1. The AC electric field of the incident wave has the form $\boldsymbol{E}(t) = \boldsymbol{E} \exp(-i\omega t) + \text{c.c.}$, where $\boldsymbol{E}$ and ω are the field amplitude and frequency, respectively, and the abbreviation "c.c." denotes the complex conjugation. The AC electric field causes the back-and-forth in-plane motion of electrons. At the edge of the structure (here, at $x = 0$), the AC motion of electrons gets distorted due to electron reflection from the edge and dynamic charge accumulation, which leads to an asymmetry of the high-frequency electron transport. This asymmetry results in the rectification of the AC current and, hence, the emergence of a DC current J_y flowing along the edge. As Figure 1 illustrates, the DC edge current can be excited by both linearly polarized and circularly polarized electromagnetic waves and the current direction is controlled by the wave polarization.

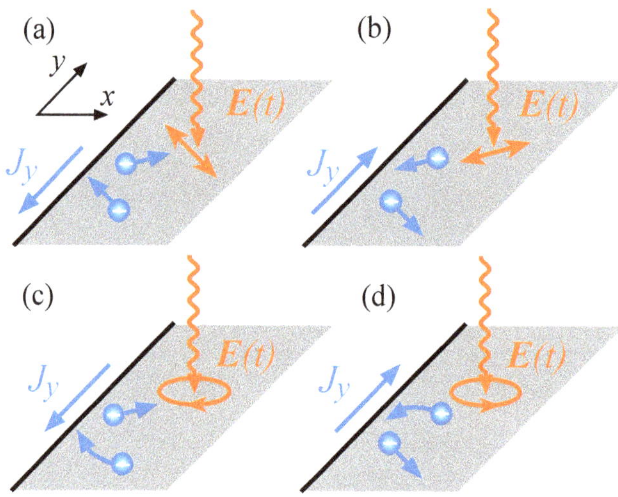

Figure 1. Illustration of the edge current formation. The back-and-forth motion of 2D carriers occupying the $x \geq 0$ half-plane by linearly polarized (**a**,**b**) or circularly polarized (**c**,**d**) AC electric field results, due to electron scattering from the edge and dynamical charge accumulation, in the DC current J_y flowing along the edge. The edge current direction is controlled by the field polarization.

Now we present the microscopic theory of high-frequency non-linear electron transport and calculate the edge currently. We consider the classical range of the electromagnetic wave frequencies, i.e., $\hbar\omega \ll E_F$, where E_F is the Fermi energy of the 2D electron gas, and describe the electron kinetics by the Boltzmann equation

$$\frac{\partial f}{\partial t} + v_x \frac{\partial f}{\partial x} + e\boldsymbol{\mathcal{E}}(x,t) \cdot \frac{\partial f}{\partial \boldsymbol{p}} = \mathrm{I}\{f\}\,. \tag{1}$$

Here, $f = f(\boldsymbol{p}, x, t)$ is the electron distribution function, $\boldsymbol{p}$ and $\varepsilon(p)$ are the electron momentum and energy, respectively, $\boldsymbol{v} = d\varepsilon/d\boldsymbol{p} = v\boldsymbol{p}/p$ is the velocity, $v = d\varepsilon/dp$, e is the electron charge, $\boldsymbol{\mathcal{E}}(x,t) = \boldsymbol{\mathcal{E}}(x)\exp(-i\omega t) + \text{c.c.}$ is the local electric field acting upon the electrons, and $\mathrm{I}\{f\}$ is the collision integral. At this stage, we assume that the electron spectrum is isotropic in the 2D plane but do not specify the exact form of the dispersion $\varepsilon(p)$. The field $\boldsymbol{\mathcal{E}}(x)$ near the edge differs from the incident field $\boldsymbol{E}$ by the correction $\delta\boldsymbol{E}(x) \parallel x$ due to the screening produced by dynamical charge accumulation [6,23,24]. Therefore, $\mathcal{E}_y = E_y$ whereas $\mathcal{E}_x(x) \propto E_x$.

At equilibrium, the electron distribution is isotropic and homogenous at $x \geq 0$ and is described by the Fermi–Dirac function $f_0(\varepsilon)$. In the presence of an AC electric field, the distribution function acquires corrections. We expand the resulting distribution function $f(\boldsymbol{p}, x, t)$ in the series in the electric field amplitude as follows

$$f(\boldsymbol{p}, x, t) = f_0 + [f_1(\boldsymbol{p}, x) \exp(-i\omega t) + \text{c.c.}] + f_2(\boldsymbol{p}, x) + \dots, \quad (2)$$

where f_1 is the first-order correction, which determines the linear (Drude) conductivity, and f_2 is the time-independent second-order correction. The second-order correction oscillating at 2ω is not considered here since it does not contribute to DC electric current.

The density of DC electric current $j_y(x)$ is determined by the asymmetric part of the correction f_2 and is given by

$$j_y(x) = eg \sum_p v_y f_2(\boldsymbol{p}, x), \quad (3)$$

where g is the factor that takes into account possible spin and valley degeneracy (e.g., $g = 2$ for GaAs quantum wells and $g = 4$ for single-layer and bilayer graphene) and $\sum_p = (2\pi\hbar)^{-2} \int d^2\boldsymbol{p}$.

Equation (1) with the perturbation term $e\boldsymbol{\mathcal{E}}(x,t) \cdot \partial f / \partial \boldsymbol{p}$ yields the following differential equations for f_1 and f_2

$$-i\omega f_1 + v_x \frac{\partial f_1}{\partial x} + e\boldsymbol{\mathcal{E}}(x) \cdot \frac{\partial f_0}{\partial \boldsymbol{p}} = \mathrm{I}\{f_1\}, \quad (4)$$

$$v_x \frac{\partial f_2}{\partial x} + \left[e\boldsymbol{\mathcal{E}}(x) \cdot \frac{\partial f_1^*}{\partial \boldsymbol{p}} + \text{c.c} \right] = \mathrm{I}\{f_2\}. \quad (5)$$

We solve Equations (4) and (5) in the approximation of elastic electron scattering in the bulk of the 2D system and for specular reflection of electrons from the edge. The latter implies that $f(p_x, p_y, 0, t) = f(-p_x, p_y, 0, t)$ which also ensures the lack of electric current across the edge. Multiplying Equation (5) by the velocity v_y and averaging the resulting equation over the directions of $\boldsymbol{p}$ one obtains

$$\left\langle v_x v_y \frac{\partial f_2}{\partial x} \right\rangle + \left\langle v_y \left(e\boldsymbol{\mathcal{E}} \cdot \frac{\partial f_1^*}{\partial \boldsymbol{p}} + \text{c.c} \right) \right\rangle = -\frac{\langle v_y f_2 \rangle}{\tau_1}, \quad (6)$$

where the angular brackets $\langle \dots \rangle$ stand for the averaging and τ_1 is the momentum relaxation time (relaxation time of the first angular harmonic) defined as $1/\tau_1 = -\langle \boldsymbol{v}\, I\{f\} \rangle / \langle \boldsymbol{v} f \rangle$. Such a definition of τ_1 enables the consideration of its dependence on the electron energy $\varepsilon(p)$. Equations (3) and (6) yield the equation for the current density

$$j_y(x) = -eg \sum_p \tau_1 v_x v_y \frac{\partial f_2}{\partial x} - eg \sum_p \tau_1 v_y \left(e\boldsymbol{\mathcal{E}} \cdot \frac{\partial f_1^*}{\partial \boldsymbol{p}} + \text{c.c} \right). \quad (7)$$

After the integration of the second term by parts, Equation (7) assumes the form

$$\begin{aligned} j_y(x) \quad &= \quad -eg \sum_p \tau_1 v_x v_y \frac{\partial f_2}{\partial x} + e^2 g \sum_p \left(\frac{\tau_1}{m} \right)' m v_x v_y (\mathcal{E}_x f_1^* + \text{c.c}) \\ &+ \quad e^2 g \sum_p \left[\frac{\tau_1}{m} + \left(\frac{\tau_1}{m} \right)' \frac{m v^2}{2} - m \left(\frac{\tau_1}{m} \right)' \frac{v_x^2 - v_y^2}{2} \right] (E_y f_1^* + \text{c.c}), \end{aligned} \quad (8)$$

where $(\dots)' = d(\dots)/d\varepsilon$ and $m = p/v = p/(d\varepsilon/dp)$ is the (energy-dependent) effective mass. In the case of parabolic dispersion $\varepsilon(p) = p^2/(2m^*)$ the effective mass $m = m^*$ is energy-independent whereas for linear dispersion $\varepsilon(p) = v_0 p$ the effective mass $m = \varepsilon/v_0^2$ linearly depends on energy. Note that the mass m also determines the quasi-classical cyclotron motion. Equation (8) is valid for arbitrary dispersion and arbitrary boundary conditions at the edge. Obviously, the DC current vanishes for the AC field polarized along or perpendicularly to the edge and emerges only if the incident field $\boldsymbol{E}$ has both x and y components. Therefore, the corrections f_1 in the second and third terms on the right-hand side of Equation (8) should be calculated for the y and x components of the field, respectively.

For specular reflection of electrons from the edge, the second sum in Equation (8) vanishes since f_1 in response to E_y is an even function of v_x. The third sum in Equation (8) can be rewritten via $\partial f_1^*/\partial x$ using the equalities

$$i\omega\langle f_1\rangle = \left\langle v_x\frac{\partial f_1}{\partial x}\right\rangle, \tag{9}$$

$$\left(i\omega-\frac{1}{\tau_2}\right)\left\langle\frac{v_x^2-v_y^2}{2}f_1\right\rangle = \left\langle v_x\frac{v_x^2-v_y^2}{2}\frac{\partial f_1}{\partial x}\right\rangle, \tag{10}$$

which follow from Equation (4). Here, τ_2 is the relaxation time of the second angular harmonic of the distribution function defined as $1/\tau_2 = -\langle(v_x^2-v_y^2)\,I\{f\}\rangle/\langle(v_x^2-v_y^2)f\rangle$. Therefore, Equation (8) assumes the form

$$\begin{aligned} j_y(x) &= -eg\sum_{p}\tau_1 v_x v_y\frac{\partial f_2}{\partial x}+\frac{e^2 g}{\omega}\sum_{p}v_x\left[\frac{\tau_1}{m}+\left(\frac{\tau_1}{m}\right)'\frac{mv^2}{2}\right]\left(iE_y\frac{\partial f_1^*}{\partial x}+\text{c.c}\right) \\ &+ e^2 g\sum_{p}mv_x\frac{v_x^2-v_y^2}{2}\left(\frac{\tau_1}{m}\right)'\left(\frac{\tau_2 E_y}{1+i\omega\tau_2}\frac{\partial f_1^*}{\partial x}+\text{c.c}\right). \end{aligned} \tag{11}$$

The current density $j_y(x)$ is determined by spatial derivatives of the distribution function and, as expected, vanishes in the 2D bulk where the electron distribution is homogenous.

The total electric current flowing along the edge is given by

$$J_y = \int_0^\infty j_y(x)dx\,. \tag{12}$$

Integrating Equation (11) over x we obtain

$$\begin{aligned} J_y &= -eg\sum_{p}\tau_1 v_x v_y[f_2(\boldsymbol{p},\infty)-f_2(\boldsymbol{p},0)] \\ &+\left\{i\frac{e^2 g}{\omega}\sum_{p}\left[\frac{\tau_1}{m}+\left(\frac{\tau_1}{m}\right)'\frac{mv^2}{2}\right]v_x[f_1^*(\boldsymbol{p},\infty)-f_1^*(\boldsymbol{p},0)]E_y+\text{c.c}\right\} \\ &+\left\{e^2 g\sum_{p}\left(\frac{\tau_1}{m}\right)'\frac{m\tau_2}{1+i\omega\tau_2}\frac{v_x^2-v_y^2}{2}v_x[f_1^*(\boldsymbol{p},\infty)-f_1^*(\boldsymbol{p},0)]E_y+\text{c.c}\right\}, \end{aligned} \tag{13}$$

where $f_n(\boldsymbol{p},0)$ and $f_n(\boldsymbol{p},\infty)$ are the corrections to the distribution function at the edge and far from the edge, respectively. For the particular case of specular reflection of electrons from the edge, the functions $f_1(\boldsymbol{p},0)$ and $f_2(\boldsymbol{p},0)$ are even in p_x and, therefore, the sums with $f_1(\boldsymbol{p},0)$ and $f_2(\boldsymbol{p},0)$ vanish. As a result, the edge current J_y is determined by the functions $f_1(\boldsymbol{p},\infty)$ and $f_2(\boldsymbol{p},\infty)$ in the bulk where the actual field $\boldsymbol{\mathcal{E}}$ coincides with the incident field $\boldsymbol{E}$. The sums with $f_1(\boldsymbol{p},\infty)$ and $f_2(\boldsymbol{p},\infty)$ can be readily calculated from Equations (4) and (5) neglecting spatial inhomogeneous terms and the electric field screening. Below, we do such calculations for the degenerate electron gas with the Fermi energy E_F.

The first-order correction in the 2D bulk has the form

$$f_1(\boldsymbol{p},\infty) = -\frac{e\tau_1 f_0'}{1-i\omega\tau_1}(\boldsymbol{v}\cdot\boldsymbol{E})\,. \tag{14}$$

Therefore, the second sum in Equation (13) is calculated as follows

$$ieg\sum_p\left[\frac{\tau_1}{m}+\left(\frac{\tau_1}{m}\right)'\frac{mv^2}{2}\right]v_x f_1^*(\boldsymbol{p},\infty)E_y+\text{c.c.}=\left[\frac{\tau_1}{m}+\left(\frac{\tau_1}{m}\right)'\frac{mv^2}{2}\right]_{E_F}[i\sigma^*(\omega)E_x^*E_y+\text{c.c.}]$$
$$=\left[\frac{\tau_1}{m}+\left(\frac{\tau_1}{m}\right)'\frac{mv^2}{2}\right]_{E_F}\text{Re}\,\sigma(\omega)(\omega\tau_1 S_2-S_3), \quad (15)$$

where $\sigma(\omega)$ is the conductivity,

$$\sigma(\omega)=-\frac{e^2 g}{2}\sum_p\frac{\tau_1 v^2 f_0'}{1-i\omega\tau_1}=\frac{ne^2}{m}\frac{\tau_1}{1-i\omega\tau_1}, \quad (16)$$

all the values are taken at the Fermi level, $n=g\sum_p f_0$ is the carrier density, and $S_2=E_xE_y^*+E_x^*E_y$ and $S_3=i(E_xE_y^*-E_x^*E_y)$ are the Stokes parameters of the incident radiation. The third sum in Equation (13) is calculated as follows

$$eg\sum_p\left(\frac{\tau_1}{m}\right)'\frac{m\tau_2}{1+i\omega\tau_2}\frac{v_x^2-v_y^2}{2}v_x f_1^*(\boldsymbol{p},\infty)E_y+\text{c.c.}=\frac{1}{4}\left(\frac{\tau_1}{m}\right)'_{E_F}\frac{mv^2\,\tau_2\,\sigma^*(\omega)}{1+i\omega\tau_2}E_x^*E_y+\text{c.c.}$$
$$=\frac{1}{4}\left(\frac{\tau_1}{m}\right)'_{E_F}\frac{mv^2\,\tau_2\,\text{Re}\,\sigma(\omega)}{1+(\omega\tau_2)^2}\left[(1-\omega^2\tau_1\tau_2)S_2+\omega(\tau_1+\tau_2)S_3\right]. \quad (17)$$

Lastly, the sum with $f_2(\boldsymbol{p},\infty)$ in Equation (13) can be expressed with the help of Equation (5) via the sum with $f_1(\boldsymbol{p},\infty)$ as follows

$$\sum_p\tau_1 v_x v_y f_2(\boldsymbol{p},\infty)=-e\sum_p\tau_1\tau_2 v_x v_y\left(\boldsymbol{E}\cdot\frac{\partial f_1^*(\infty)}{\partial\boldsymbol{p}}+\text{c.c}\right). \quad (18)$$

Integration of the right-hand side of Equation (18) by parts gives

$$\sum_p\tau_1 v_x v_y f_2(\boldsymbol{p},\infty)=\left\{e\sum_p\left[\frac{\tau_1\tau_2}{m}+\frac{m^2v^2}{2}\left(\frac{\tau_1\tau_2}{m^2}\right)'\right](v_yE_x^*+v_xE_y^*)f_1(\boldsymbol{p},\infty)+\text{c.c.}\right\}$$
$$+\left\{e\sum_p m^2\left(\frac{\tau_1\tau_2}{m^2}\right)'\frac{v_x^2-v_y^2}{2}(v_yE_x^*-v_xE_y^*)f_1(\boldsymbol{p},\infty)\right\}. \quad (19)$$

The above sums can be calculated similarly to the sums in Equations (15) and (17), which yields

$$g\sum_p\tau_1 v_x v_y f_2(\boldsymbol{p},\infty)=2\left[\frac{\tau_1\tau_2}{m}+\frac{m^2v^2}{4}\left(\frac{\tau_1\tau_2}{m^2}\right)'\right]_{E_F}\text{Re}\,\sigma(\omega)S_2. \quad (20)$$

Finally, summing up all contributions to the edge current we obtain

$$J_y=\frac{e\text{Re}\,\sigma(\omega)}{m}\left\{\tau_1(\tau_1-2\tau_2)+\frac{m^2v^2}{2}\left[\frac{\tau_1}{2}\left(\frac{\tau_1}{m}\right)'-m\left(\frac{\tau_1\tau_2}{m^2}\right)'\right]+\frac{m^2v^2(\tau_1+\tau_2)}{4[1+(\omega\tau_2)^2]}\left(\frac{\tau_1}{m}\right)'\right\}S_2$$
$$-\frac{e\text{Re}\,\sigma(\omega)}{m\omega}\left[\tau_1+\frac{m^2v^2[2+\omega^2\tau_2(\tau_2-\tau_1)]}{4[1+(\omega\tau_2)^2]}\left(\frac{\tau_1}{m}\right)'\right]S_3. \quad (21)$$

Equation (21) represents the main result of this paper. It describes the DC edge current in 2D electron gas with an arbitrary electron dispersion $\varepsilon(p)$ and arbitrary energy-dependent relaxation times.

For parabolic energy spectrum with $\varepsilon(p) = p^2/(2m^*)$, Equation (21) gives

$$J_y^{(\mathrm{par})} = \frac{e\mathrm{Re}\,\sigma(\omega)}{m^*}\left\{\tau_1(\tau_1 - 2\tau_2) + \left[\frac{\tau_1\tau_1'}{2} - (\tau_1\tau_2)'\right]\varepsilon_F + \frac{(\tau_1+\tau_2)\tau_1'\varepsilon_F}{2[1+(\omega\tau_2)^2]}\right\}S_2 - \frac{e\mathrm{Re}\,\sigma(\omega)}{m^*\omega}\left\{\tau_1 + \frac{2+\omega^2\tau_2(\tau_2-\tau_1)}{2(1+\omega^2\tau_2^2)}\tau_1'\varepsilon_F\right\}S_3\,. \quad (22)$$

This result was obtained previously in the approximation of a single energy-independent relaxation time ($\tau_1 = \tau_2 = \mathrm{const}$) in Ref. [6] and in the form of Equation (22) in Ref. [8].

For linear energy spectrum $\varepsilon(p) = v_0 p$, Equation (21) gives

$$J_y^{(\mathrm{lin})} = \frac{ev_0^2\,\mathrm{Re}\,\sigma(\omega)}{2\varepsilon_F}\left\{\tau_1\left(\frac{3\tau_1}{2} - 2\tau_2\right) + \left[\frac{\tau_1\tau_1'}{2} - (\tau_1\tau_2)'\right]\varepsilon_F + \frac{(\tau_1+\tau_2)(\tau_1'\varepsilon_F - \tau_1)}{2[1+(\omega\tau_2)^2]}\right\}S_2 - \frac{ev_0^2\,\mathrm{Re}\,\sigma(\omega)}{2\varepsilon_F\omega}\left\{2\tau_1 + \frac{[2+\omega^2\tau_2(\tau_2-\tau_1)](\tau_1'\varepsilon_F-\tau_1)}{2[1+(\omega\tau_2)^2]}\right\}S_3\,. \quad (23)$$

3. Results and Discussion

Now, we discuss the spectral and polarization dependence of the edge current in 2D systems with parabolic and linear dispersions, also for different scattering mechanisms. First, we note that the current J_y is proportional to the square of the electric field amplitude ($S_2, S_3 \propto E^2$), i.e., to the intensity of the incident field. Therefore, it belongs to the class of photocurrents. Here, the photocurrent emerges due to the intraband (Drude-like) absorption of the electromagnetic field and is proportional to the real part of the high-frequency conductivity $\sigma(\omega)$.

The direction of the edge current (the polarity along the y axis) depends on the polarization state of the field via the Stokes parameters $S_2 = (E_xE_y^* + E_yE_x^*)$ and $S_3 = i(E_xE_y^* - E_yE_x^*)$. The contribution $J_y \propto S_2$ is induced by linearly polarized radiation. This current is maximal if the radiation is polarized at the angle $\pm\pi/4$ with respect to the edge and vanishes if the radiation is polarized along or perpendicular to the edge, Figure 1a,b. The contribution $J_y \propto S_3$ describes the edge current induced by circularly polarized radiation and its direction is controlled by the radiation helicity, Figure 1c,d.

Equation (21) is quite general and can be applied to a wide class of 2D systems, including conventional III-V and II-VI quantum wells, bilayer graphene, and transition metal dichalcogenide monolayers with the parabolic spectrum, graphene and HgTe/CdHgTe quantum wells of the critical thickness with the linear spectrum, and narrow-gap 2D systems with the Dirac-like spectrum $\varepsilon(p) = v_0\sqrt{(m^*v_0)^2 + p^2}$. Below, we calculate J_y for single-layer and bilayer graphene.

Figure 2 shows the frequency dependence of the edge photocurrent J_y in a 2D system with the linear dispersion law $\varepsilon(p) = v_0 p$. The parameters used for calculations are given in the caption of Figure 2 and correspond to high-quality graphene [25] with the electron density $n = 5 \times 10^{11}$ cm^{-2}. The curves are calculated after Equation (23) for linearly polarized radiation with the electric field directed at the angle $\pi/4$ with respect to the edge ($S_2/E^2 = 1, S_3 = 0$) and for circularly polarized radiation ($S_3/E^2 = 1, S_2 = 0$). We consider two model types of scattering potential: (i) short-range scatterers, resulting in the energy dependence of the relaxation times $\tau_1 = 2\tau_2 \propto \varepsilon^{-1}$, and (ii) charged scatterers with the Coulomb potential, resulting in $\tau_1 = 3\tau_2 \propto \varepsilon$ [7]. For charged scatterers, the ratio $\tau_{1,2}/m$ does not depend on energy and the edge current is given by $J_y = e\mathrm{Re}\sigma(\omega)v_0^2\tau_1^2(E_F)/3E_F$ for linearly polarized radiation and $J_y = -e\mathrm{Re}\sigma(\omega)v_0^2\tau_1(E_F)/\omega E_F$ for circularly polarized radiation. Hence, the frequency dependence of the edge currents excited by linearly and circularly polarized radiation are determined by $\mathrm{Re}\sigma(\omega) \propto 1/(1+\omega^2\tau_1^2)$ and $\mathrm{Re}\sigma(\omega)/\omega$, respectively. For short-range scatterers, all terms in Equation (21) contribute to the current, and the frequency dependence gets more complicated. In particular, the current induced by linearly polarized radiation is constant at low frequencies, changes its sign at an intermediate frequency, and decays as $\propto \omega^{-2}$ at high frequencies. The circular contribution for short-range scatterers behaves as $\propto \omega$ at low frequencies, in contrast to the diverging behavior $\propto \omega^{-1}$ for Coulomb scatterers. The current magnitude is $J_y \sim 10$ nA for 1 W/cm^2 of the radiation intensity.

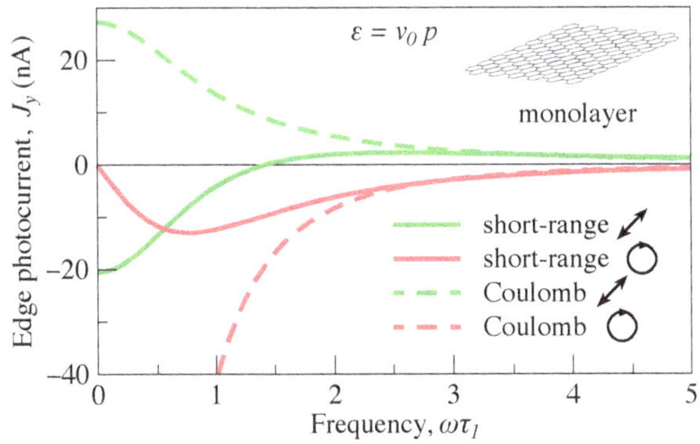

Figure 2. Frequency dependence of the edge current in 2D systems with a linear dispersion of carriers. Solid curves correspond to short-range scatterers and linearly polarized field with $S_2 = E^2$ (green curve) and circularly polarized field with $S_3 = E^2$ (red curve). Dashed curves present the results for scattering by charged impurities. The curves are calculated after Equation (23) for the parameters corresponding to single-layer graphene: $v_0 = 10^8$ cm/s, $n = 5 \times 10^{11}$ cm^{-2} (the Fermi energy $E_F \approx 80$ meV and the effective mass at the Fermi level $m \approx 0.013\, m_0$), $\tau_1(E_F) = 1$ ps, and $E = 8$ V/cm corresponding to the radiation intensity $I = 1$ W/cm^2.

Figure 3 shows the frequency dependence of the edge photocurrent J_y in the 2D system with the parabolic dispersion law $\varepsilon(p) = p^2/2m^*$. The parameters used for calculations are given in the caption of Figure 3 and correspond to high-quality bilayer graphene [6]. We use the same electron density $n = 5 \times 10^{11}$ cm^{-2} as in the case of single-layer graphene. The curves are calculated after Equation (22) for linearly polarized radiation with the electric field directed at the angle $\pi/4$ with respect to the edge ($S_2/E^2 = 1$, $S_3 = 0$) and for circularly polarized radiation ($S_3/E^2 = 1$, $S_2 = 0$). For 2D systems with the parabolic spectrum and short-range scatterers, both the mass m and relaxation times $\tau_1 = \tau_2$ are independent of energy and Equation (21) yields $J_y = -e\mathrm{Re}\sigma(\omega)\tau_1^2(E_F)/m^*$ for linearly polarized radiation and $J_y = -e\mathrm{Re}\sigma(\omega)\tau_1(E_F)/\omega m^*$ for circularly polarized radiation. For Coulomb scatterers, $\tau_1 = 2\tau_2 \propto \varepsilon$, and the frequency dependence of J_y is more complicated. Note that both the direction and magnitude of the current are determined to a great extent by the scattering mechanism. The calculated current magnitude for bilayer graphene is of the order of several nA for the radiation intensity 1 W/cm^2 and is slightly smaller than that for monolayer graphene at the same electron density due to the larger effective mass in bilayer graphene.

Experimentally, edge photocurrents are detected as electric currents in short circuits or as voltage drops between contacts in open circuits [5,6,9,10]. If the sample is small and fully illuminated, the photocurrents are generated along all the edges and the resulting distribution of the photoinduced electric potential in the sample is determined by the edge photocurrents and the compensating drift currents in the sample. For linearly polarized radiation, the photocurrents generated along different edges are generally not equal because of different orientations of the edges with respect to the electric field polarization. For circularly polarized radiation, the edge photocurrents form a vortex whose winding direction depends on the radiation helicity. The photocurrents circulating around the sample produce, in turn, a magnetic field, which can be seen as a manifestation of the inverse Faraday effect.

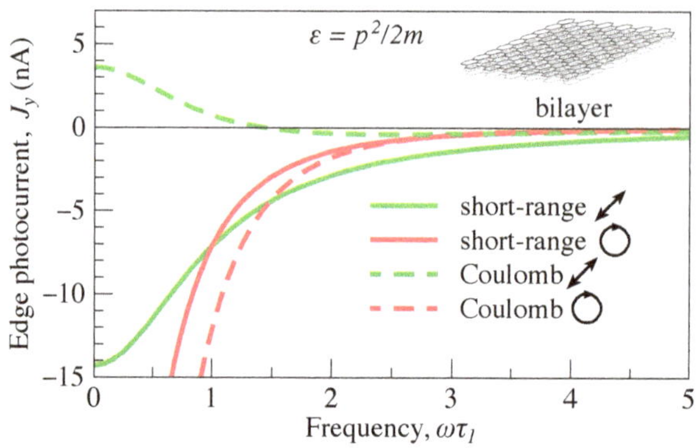

Figure 3. Frequency dependence of the edge photocurrent in 2D systems with parabolic energy dispersion. Solid curves correspond to short-range scatterers and linearly polarized field with $S_2 = E^2$ (green curve) and circularly polarized field with $S_3 = E^2$ (red curve). Dashed curves present the results for scattering by charged impurities. The curves are calculated after Equation (22) for parameters corresponding to bilayer graphene: $m^* = 0.03\, m_0$, $n = 5 \times 10^{11}$ cm^{-2}, $\tau_1(E_F) = 1$ ps, and $E = 8$ V/cm corresponding to the radiation intensity $I = 1$ W/cm^2.

4. Conclusions

To conclude, we have developed a kinetic theory of the edge photogalvanic effect for the intraband electron transport in two-dimensional materials. It is shown that the back-and-forth motion of charge carriers by AC electric field of incident radiation is distorted at the edges of the sample resulting in direct electric currents flowing along the edges. The edge current direction is controlled by the radiation polarization while its spectral dependence is determined by the carrier dispersion and the mechanism of carrier scattering. We have obtained an analytical expression for the edge current valid for arbitrary dispersion law and scattering mechanism and analyzed the result for single-layer and bilayer graphene for electron scattering by short-range defects and Coulomb impurities. Considering the important role of edge regions in small-size samples such as flakes of two-dimensional crystals, one can expect that the edge photogalvanic effect will find applications in fast detectors of terahertz radiation and radiation polarization.

Author Contributions: M.V.D. and S.A.T. have contributed equality to the paper. All authors have read and agreed to the published version of the manuscript.

Funding: This research was supported by the Russian Science Foundation (project 21-72-00047).

Institutional Review Board Statement: Not applicable.

Informed Consent Statement: Not applicable.

Data Availability Statement: Not applicable.

Conflicts of Interest: The authors declare no conflict of interest.

Abbreviations

The following abbreviations are used in this manuscript:

2D Two-dimensional

References

1. Novoselov, K.S.; Jiang, D.; Schedin, F.; Booth, T.J.; Khotkevich, V.V.; Morozov, S.V.; Geim, A.K. Two-dimensional atomic crystals. *Proc. Natl. Acad. Sci. USA* **2005**, *102*, 10451. [CrossRef]
2. Geim, A.K.; Grigorieva, I.V. Van der Waals heterostructures. *Nature* **2013**, *499*, 419. [CrossRef] [PubMed]

3. Koppens, F.H.L.; Mueller, T.; Avouris, P.; Ferrari, A.C.; Vitiello, M.S.; Polini, M. Photodetectors based on graphene, other two-dimensional materials and hybrid systems. *Nat. Nanotechnol.* **2014**, *9*, 780. [CrossRef] [PubMed]
4. Bandurin, D.A.; Svintsov, D.; Gayduchenko, I.; Xu, S.G.; Principi, A.; Moskotin, M.; Tretyakov, I.; Yagodkin, D.; Zhukov, S.; Taniguchi, T.; et al. Resonant terahertz detection using graphene plasmons. *Nat. Commun.* **2018**, *9*, 5392. [CrossRef]
5. Karch, J.; Drexler, C.; Olbrich, P.; Fehrenbacher, M.; Hirmer, M.; Glazov, M.M.; Tarasenko, S.A.; Ivchenko, E.L.; Birkner, B.; Eroms, J.; et al. Terahertz Radiation Driven Chiral Edge Currents in Graphene. *Phys. Rev. Lett.* **2011**, *107*, 276601. [CrossRef] [PubMed]
6. Candussio, S.; Durnev, M.V.; Tarasenko, S.A.; Yin, J.; Keil, J.; Yang, Y.; Son, S.K.; Mishchenko, A.; Plank, H.; Bel'kov, V.V.; et al. Edge photocurrent driven by terahertz electric field in bilayer graphene. *Phys. Rev. B* **2020**, *102*, 045406. [CrossRef]
7. Glazov, M.; Ganichev, S. High frequency electric field induced nonlinear effects in graphene. *Phys. Rep.* **2014**, *535*, 101. [CrossRef]
8. Durnev, M.V.; Tarasenko, S.A. Rectification of AC Electric Current at the Edge of 2D Electron Gas. *Phys. Status Solidi (B)* **2021**, *258*, 2000291. [CrossRef]
9. Candussio, S.; Durnev, M.V.; Slizovskiy, S.; Jötten, T.; Keil, J.; Bel'kov, V.V.; Yin, J.; Yang, Y.; Son, S.K.; Mishchenko, A.; et al. Edge photocurrent in bilayer graphene due to inter-Landau-level transitions. *Phys. Rev. B* **2021**, *103*, 125408. [CrossRef]
10. Candussio, S.; Golub, L.E.; Bernreuter, S.; Jötten, T.; Rockinger, T.; Watanabe, K.; Taniguchi, T.; Eroms, J.; Weiss, D.; Ganichev, S.D. Nonlinear intensity dependence of edge photocurrents in graphene induced by terahertz radiation. *Phys. Rev. B* **2021**, *104*, 155404. [CrossRef]
11. Durnev, M.V.; Tarasenko, S.A. Edge photogalvanic effect caused by optical alignment of carrier momenta in two-dimensional Dirac materials. *Phys. Rev. B* **2021**, *103*, 165411. [CrossRef]
12. Yin, X.; Ye, Z.; Chenet, D.A.; Ye, Y.; O'Brien, K.; Hone, J.C.; Zhang, X. Edge Nonlinear Optics on a MoS2 Atomic Monolayer. *Science* **2014**, *344*, 488. [CrossRef] [PubMed]
13. Mishina, E.D.; Sherstyuk, N.E.; Shestakova, A.P.; Lavrov, S.D.; Semin, S.V.; Sigov, A.S.; Mitioglu, A.; Anghel, S.; Kulyuk, L. Edge effects in second-harmonic generation in nanoscale layers of transition-metal dichalcogenides. *Semiconductors* **2015**, *49*, 791. [CrossRef]
14. Durnev, M.V.; Tarasenko, S.A. Second harmonic generation at the edge of a two-dimensional electron gas. *Phys. Rev. B* **2022**, *106*, 125426. [CrossRef]
15. Plank, H.; Durnev, M.V.; Candussio, S.; Pernul, J.; Dantscher, K.M.; Mönch, E.; Sandner, A.; Eroms, J.; Weiss, D.; Bel'kov, V.V.; et al. Edge currents driven by terahertz radiation in graphene in quantum Hall regime. *2D Mater.* **2019**, *6*, 011002. [CrossRef]
16. Magarill, L.I.; Entin, M.V. Photogalvanic effect in films. *Sov. Phys. Solid State* **1979**, *21*, 743.
17. Alperovich, V.L.; Belinicher, V.I.; Novikov, V.N.; Terekhov, A.S. Surface photovoltaic effect in solids. Theory and experiment for interband transitions in gallium arsenide. *Sov. Phys. JETP* **1981**, *53*, 1201.
18. Alperovich, V.L.; Minaev, A.O.; Terekhov, A.S. Ballistic electron transport through epitaxial GaAs films in a magneti- cally induced surface photocurrent. *JETP Lett.* **1989**, *49*, 702.
19. Gurevich, V.L.; Laiho, R. Photomagnetism of metals: Microscopic theory of the photoinduced surface current. *Phys. Rev. B* **1993**, *48*, 8307–8316. [CrossRef]
20. Schmidt, C.B.; Priyadarshi, S.; Tarasenko, S.A.; Bieler, M. Ultrafast magneto-photocurrents in GaAs: Separation of surface and bulk contributions. *Appl. Phys. Lett.* **2015**, *106*, 142108. [CrossRef]
21. Mikheev, G.M.; Saushin, A.S.; Styapshin, V.M.; Svirko, Y.P. Interplay of the photon drag and the surface photogalvanic effects in the metal-semiconductor nanocomposite. *Sci. Rep.* **2018**, *8*, 8644. [CrossRef] [PubMed]
22. Mikheev, G.M.; Fateev, A.E.; Kogai, V.Y.; Mogileva, T.N.; Vanyukov, V.V.; Svirko, Y.P. Helicity dependent temporal profile of the semiconductor thin film photoresponse. *Appl. Phys. Lett.* **2021**, *118*, 201105. [CrossRef]
23. Volkov, V.; Mikhailov, S. Edge magnetoplasmons: Low frequency weakly damped excitations in inhomogeneous two-dimensional electron system. *JETP* **1988**, *67*, 1639.
24. Zabolotnykh, A.A.; Volkov, V.A. Interaction of gated and ungated plasmons in two-dimensional electron systems. *Phys. Rev. B* **2019**, *99*, 165304. [CrossRef]
25. Castro Neto, A.H.; Guinea, F.; Peres, N.M.R.; Novoselov, K.S.; Geim, A.K. The electronic properties of graphene. *Rev. Mod. Phys.* **2009**, *81*, 109. [CrossRef]

MDPI
St. Alban-Anlage 66
4052 Basel
Switzerland
www.mdpi.com

Applied Sciences Editorial Office
E-mail: applsci@mdpi.com
www.mdpi.com/journal/applsci

www.ingramcontent.com/pod-product-compliance
Lightning Source LLC
Chambersburg PA
CBHW041951260926
51850CB00019B/2621
* 9 7 8 3 0 3 6 5 9 3 7 0 8 *